CW01510906

Interrupted Journeys

Interrupted Journeys

Badgers and Other Roadside Distractions

ADRIAN POTTER

JOHN MURRAY

First published in Great Britain in 2025 by John Murray (Publishers)

1

Copyright © Adrian Potter 2025

The right of Adrian Potter to be identified as the Author
of the Work has been asserted by him in accordance with
the Copyright, Designs and Patents Act 1988.

A CIP catalogue record for this title is available from the British Library

Hardback ISBN 9781399822404
ebook ISBN 9781399822428

Typeset in Bembo by Hewer Text UK Ltd, Edinburgh
Printed and bound in Great Britain by Clays Ltd, Elcograf S.p.A.

John Murray policy is to use papers that are natural, renewable and
recyclable products and made from wood grown in sustainable forests.
The logging and manufacturing processes are expected to conform
to the environmental regulations of the country of origin.

Carmelite House
50 Victoria Embankment
London EC4Y 0DZ

www.johnmurraypress.co.uk

John Murray Press, part of Hodder & Stoughton Limited
An Hachette UK company

The authorised representative in the EEA is Hachette Ireland, 8 Castlecourt
Centre, Dublin 15, D15 XTP3, Ireland (email: info@hbgi.ie)

Contents

This then is my story . . . It has bits of marrow sticking to it, and blood, and beautiful bright-green flies.

Vladimir Nabokov, *Lolita* (1955)

Author's Note

The names of almost all the people (and even badgers!) in what follows have been changed. Place names have largely been avoided. At no point was the internet consulted in the writing of this book.

Prologue

A BADGER IS NOCTURNAL. It is short-legged and low to the ground. As far as possible, it keeps to cover. It is not easy to observe, either in the dark or in undergrowth. Neither can it *see* a great deal itself. A deer stands tall, its head on a long neck. Its large eyes and ears can usefully detect distant information from a high vantage point. It makes much more sense for a badger to rely on its nose. It can't *see* where it's going, it *smells* where it's going.

Compared with a fox, too, a badger's eyes and ears are small. Its eyes are 'piggy', dark and inscrutable, its white-edged ears are blunt rather than pointed like a fox's. Deer and fox leave scent signals via glands in their feet. This adaptation eliminates the need for these taller animals to squat down and apply scent, which would demand a lot of wasted energy. Scent glands on other parts of the body are used to anoint vegetation and other features of the environment at a variety of levels. Wild mammals pass along olfactory tunnels redolent with reassuring and enticing smells.

Badgers are social animals, living in colonies, which makes them unusual among British carnivores. They need to be able to recognise one another, friend or foe. Their bold black-and-white masks enable them to be seen by each other in low light at close quarters. Clan members share the same unique smell and regularly 'musk', rubbing each others' pelts with scent from a subcaudal gland located beneath their tail.

Badgers can recognise intruders from neighbouring clans just as they can identify kin, by smell.

Wild animals follow regular trails in their daily, or nightly, routines. Badgers in particular create well-worn paths through the undergrowth. These are efficient routes along which they pass to their foraging grounds and other places of significance. Being low-slung, it is little effort and takes only the briefest of pauses for a badger to squat down and scent mark while on the move, refreshing its trail. Where a badger path leads under a low strand of barbed wire, distinctive hairs (long and coarse; coloured white, black, white) snag on the barbs, building up into telltale tufts.

Badger paths often, inevitably, issue on to roads. On its own network of familiar highways a badger feels secure and travels confidently at a jaunty pace. But when it comes to a man-made road its path is abruptly broken and its sense of security evaporates; the reassuring scent trail has mysteriously run out, as has the concealing cover. The animal pauses in the verge and raises its snout to sniff the air for danger, as it would when cautiously thinking about emerging from its burrow after sunset. The smells and sounds it now detects are bewildering, especially those strange sounds that are hard to judge the distance of since what is making them is travelling so fast.

It is late and the traffic is intermittent now. Once on the tarmac the badger is exposed and feels vulnerable, and so it will run in a panic – but it is hesitant at first. As it edges uncertainly forward it is suddenly dazzled and freezes in a defensive posture, hair fluffed out to increase its size in the eyes of an enemy. But this enemy is implacable, and musk is released in a reflex response to fear and alarm . . .

I

The Clough

YORKSHIREMEN – AND WOMEN – have always been wary of comers-in. A new deputy head at our school from 'down South' was overheard to pronounce that particular local landscape feature known as a clough so that it rhymed not with 'tough' and 'enough' but with 'plough' and 'bough'. Was it possible that he had never heard of Brian Clough? For some reason or other, perhaps because the forward-thinking deputy made himself rather unpopular with the old guard, this trivial faux pas passed into staffroom folklore along with the time Virginia, who later, appointed from within, would become a well-liked deputy, turned up for work wearing 'pretty' silver earrings which I helpfully pointed out were in the shape of cannabis leaves, a fact that certainly wouldn't have been lost on some of the children.

At the age of fifty-three – still young in some envious people's eyes, certainly still able-bodied – I had worked at the same junior school for twenty-five years and never missed a day. I had worked under four different heads and taught the children of the children I'd taught. I suddenly felt I had had enough – no, it had been coming – and I was out of touch with the new technology. Interactive whiteboards weren't for me. My proud copperplate handwriting on blackboard or wipe-clean whiteboard was obsolete. I took early retirement. Such decisions are always tough at the time, but I had no dependants and had always lived frugally in my solitude. When I was sixty I could begin to claim my annual teachers' pension and would receive a lump sum in addition. In the meantime I had savings; I was financially secure. I had an idea that I would go 'feral' and also that I wanted to be active in conservation, as a volunteer.

I lived in a river valley system which might have come straight out of a geography textbook, it was that discrete and well defined, an offshoot of the rest of the borough with its own distinct natural and human histories. My terraced house backed on to the river. The river was the basis for all the valley's textile mills, like the empty, blank-windowed building on the opposite bank. The house had no garden. The birds I saw each day from my back windows were kingfishers and herons, dippers and grey wagtails.

From the big kitchen window I could see a large flat stone jutting out of the water which was only submerged when the river was high and roiling. Here birds – mostly pigeons – came down to drink. It was a platform from which the wagtails launched themselves, like self-propelled shuttlecocks on elastic, to catch insects in mid-air; it was a stopping-post for the dippers who caught their prey underwater; and once a sandpiper was there, a summer migrant from Africa, perched on the stone, bobbing, close enough for me to photograph. I fed the fish – brown trout – by dropping pellets of bread crust from the open window. I caught glimpses of their iridescent flanks as they rose and flipped over quick as ricocheting projectiles. Sometimes I saw mink, which is always a thrill despite what anyone says. Often I dreamt of seeing an otter.

I bought a pair of waders from a fishing-tackle shop so I could remove large unsightly items of rubbish from the river: a wheelie bin, traffic cones, a bicycle frame, sun canopies lifted by the wind from the wine bar upstream. The lesser items sailed away towards town. But the water was clear, if peaty. At this point the terraced row and the derelict mill opposite canalised the river so that the water lapped against the walls and left a dirty tidemark. A few trees – sycamores – had still managed to gain a foothold. The clear airspace over the water was a flight corridor for birds and a canny sparrow-hawk perched in the branches, waiting in ambush. In the evenings, bats dipped and looped. My front door opened on to the street. There was nothing of interest to see in that direction apart from the antics of the jackdaws.

Actually, the jackdaws could be very entertaining. There was the time a fledged chick with a mouth as wide as a frog's took refuge behind one of my wheelie bins – this dark space probably resembled the cavity it occupied as a nestling – and the two parent birds perched protectively on the stone lintel above my front door, adjacent to the bins, launching noisy assaults on me as I exited and entered my house. And the occasion when one perched expectantly on the gutter of the Indian restaurant opposite waiting for the emergence of fledgling starlings from beneath the eaves, ignoring the empty threats and insults of the frantic parent starlings. I knew what was going to happen but there was nothing I could do about it and, besides, it wasn't my place to interfere. So I just watched. And when the boldest feathered chick sallied forth into the world, full of the spirit of adventure, it was immediately stabbed bloodily to death.

The water itself was endlessly fascinating, in different lights, under different weather conditions, at different levels and rates of flow. There would be a moment during each day when I found myself looking out and down, letting my thoughts and attention drift. There comes the idle whimsy when one is lucky enough to live alongside a river that sooner or later the whole world in one form or another will come by, that all things will pass; a sense of continuity and longevity. The river never stopped flowing. It rose dramatically after heavy rains with the run-off from the surrounding moors and hills and was turbulent and turbid, but it soon settled down again. I always wondered how the aquatic creatures coped; whether they managed to cling on or were swept away downstream in periodic shunts.

When I did capture an otter on trail camera, it was in a rather different, humdrum setting, under a road bridge on the outskirts of town. I was monitoring badger and fox traffic along the bank of a brook when the otter showed up unexpectedly. It appeared twice, ten days apart. Otters are great wanderers with long, linear territories, and waiting for one to come by requires great patience.

Along the length of my home river on both sides, many streams feed into the deeper channel, draining off the high ground north and south. Each stream or brook is associated with its own lesser valley, or clough, more or less perpendicular to the west–east course of the main valley. These tend to be narrow and steep in more than one axis. Because of the rugged nature of the terrain they have been spared the axe and the plough, and many are wooded. Some have delightful names. My personal favourite is Purl Clough, which, apart from evoking the preciousness of its homophone, has a charming derivation, being a corruption or contraction of Purrill (Purr-Rill), the rill that purrs. These cloughs are important refuges and corridors for wildlife. Their slopes make excellent sites for well-drained badger setts bolstered and defended by tree roots. Deer lie up under cover of the foliage during the day, their presence often unsuspected.

Progressive human beings aren't likely to miss a trick, however. Cloughs were seen as convenient places to dump the megatons of industrial- and domestic-grade ash which were the excretions of the mills and their dependants. A so-called bottle tip is where a local dump of domestic waste includes plate and glassware among the ash, some of it likely to be in pristine condition and of some value. Happily, ash tips are not so detrimental to wildlife. Badgers actually seem to favour the friable substrate for its ease of digging and drainability. Responsible – and law-abiding – treasure hunters ('bottle-diggers') steer well clear of tip sites occupied by badgers, but badger enthusiasts who monitor setts may be rewarded with the spoils of their subjects' labour. To a badger burrowing in ash, a Victorian or Edwardian bottle with raised lettering in coloured glass is no more or less than a mineral lump akin to a stone to be winkled out of the ground and dragged on to its own waste heap. Badgers are accidental archaeologists unearthing antique artefacts as if from a bran tub and casting them aside for others to pick over.

Donning rose-tinted spectacles we look nostalgically upon all this old *rubbish* as being quaint and of historical interest, carelessly forgetting that it is just a load of rubbish and that we have always been a bunch of tossers. Today's rubbish is so *ugly* and vulgar by comparison and it's often dumped in beautiful places. Fly-tipping is, by definition, carried out well away from the point of origin, out of sight out of mind. The shameless tosser doesn't care about the wood, ravine or whatever he chooses to dump his load in, whether it's cannabis-growing surplus and paraphernalia or the erstwhile contents of a gutted bathroom or kitchen. The main thing is just to get rid of it all without incurring a charge. Those warning notices with their empty threats aren't fooling anyone. Yet, still the wildlife isn't discouraged by the aesthetics of the post-apocalypse. *Still* we will find badgers here like thick-skinned, laid-back Wombles. *Brock* or *bruck* is a Scots word for rubbish; a brock is 'a dirty, stinking fellow' (*Chambers Dictionary*).

A certain clough at the head of the valley hasn't been honoured with a name. On the Ordnance Survey (OS) map, clearly out of date, it isn't even marked by any tree symbols, yet it is well wooded. The few original big trees are ash and sycamore. Their mossy trunks and lower limbs lean out over the rocky stream with its series of modest waterfalls. Long-disused rope swings, also covered in moss, hang down like lianas. The landowners had the foresight to plant a wide range of extra broad-leaved trees forty years ago, the most successful of which has been that water-loving species alder. The clough is on private property, fenced with sheep mesh along its margins. Walkers taking the footpath that cuts through it on to the moor are greeted by a sign: CAUTION STEEP GROUND. Few would guess that it is host to a long-established badger clan, that its vegetation conceals one of the most active and impressive setts in the valley, with a broad scree slope formed by the excavations of many generations of burrowers.

The stream has its own resident dippers and the damp level ground attracts woodcock in winter. The stream pays no heed to summer or winter but keeps on flowing, through long droughts when the badgers are hard-pressed to find earthworms to eat, and bitter cold spells when icicles as long and thick as a man's arm hang from the sheer sides.

In the grand scheme of things – to a person passing through breadthways – it might not seem like much, but to one who straddles the mesh and discovers the sett and takes the time to look around, to one who comes to know it long-term as a pocket of tranquillity where others aren't likely to intrude and values it as the last tree cover before the open moor as well as perennial home base to large nocturnal mammals; to them the clough takes on great local significance. In 2020, the year I turned sixty (and a memorable year for all of us), I could say that this special place belonged to me. What's more, the clough was just part of a wider plot of land that came into my possession. How did I get here? First came the RSPB, then the badger group. The two had overlapped.

Here in this valley, we are surrounded by the moors. It may be imagination but they seem to call to me, to let me know somehow that they will always be there when I need them. Throughout my teaching career I sought them out at weekends and during holidays. I found their peace and wild, elemental nature inviting and therapeutic. The wide open spaces and open access, the lack of people or signs of human- ity; these things attracted me. The sparseness of points of interest brought each feature and creature encountered into clearer focus. It was natural, therefore, in my professional retirement that I should be interested in what the RSPB were up to, specifically their long-term project of blanket bog restoration. The moors were drying out and the peat turning to dust, which wasn't good for wildlife, for biodiver- sity, or for the environment or climate. The moors are riddled with gulleys. By blocking these channels to slow the flow of water and encouraging sphagnum moss to grow, the dry, bare peat could become well-vegetated bog as it once was. Sphagnum is called an 'ecosystem engineer' for its knock-on beneficial effects on other organisms; it holds water like a sponge and gives other plants a chance to take hold. These conditions are good for insects which feed breeding waders such as golden plover, dunlin and curlew, species in worrying decline, which is where the RSPB come in. I wasn't the only outsider interested in this process. The work was well funded because another of its positive outcomes is carbon capture. This was cutting-edge conser- vation and I wanted in.

It was tough, physical work, but all the more satisfying for that. Sometimes – often – it was so cold on the exposed tops outside the breeding season that such demanding labour was the only way to keep the blood circulating. Just as hay fever is the bane of my summers, swollen chilblained fingers became the scourge of my winters.

For six years, two days a week, I was an RSPB volunteer. For much of the rest of the time, including evenings – the best time of day – I explored my immediate environs in depth, getting to know better the reservoirs, woods, cloughs and hidden places, beginning to learn their secrets. If I was ever caught trespassing I could say I was with the RSPB. At some point I crossed paths with the chair of the badger group out and about, and realised from our conversation that here was another way, as with the RSPB, of legitimately getting on the inside and gaining access to further, deeper secrets. I didn't know a great deal about badgers then; I knew where a handful of setts were and had noticed – sometimes stopping to photograph – a few road victims on my travels.

Soon after joining the group it became clear that very few of its members were active in the field. Because of my credentials as a conservation volunteer, I was entrusted with a photocopy of a section of OS map with asterisks marking known sett locations. I was surprised at how many there were and set out to track down as many as I could. 'Badgering' is highly addictive. A series of cunningly placed clues leads to the solving of a mystery. More than any other British animal, the badger leaves a wide range of distinctive traces of its unseen activities. And tracking a wild beast to its lair awakens latent primal urges. At another level, it is like being a child again, getting dirty and playing at being commandos or explorers. Best of all were the times when I could announce that I had found a sett which wasn't already marked on the map.

Given more and more responsibilities within the group, after a year I was asked if I would take on the 'Badger Phone', which seemed to be something of a hot potato, changing hands from one month to the next. This was to be, believe it or not, my first experience of a mobile phone (the year was

2017), a bottom-of-the-range push-button Alcatel which became – and still is – my personal phone. Around the same time I was also given responsibility for casualties, first of all in my own political ward but very soon in the whole borough – one of the largest in the country. The group's records showed that responsibility for casualties had been somewhat neglected, but it seemed to me that this ought to be a priority public service since a dead animal at the roadside presents the commonest type of badger encounter. I also sensed an exciting opportunity to amass my own raw data, and I duly disseminated the incident number far and wide. I became the go-to man for the council (highways and cleansing departments), police, vets and general public. I also took on foxes and would later add deer. I was so busy that eventually I could no longer spare two days on the moors, which was a pity. I was on call morning to night every day of the year.

It was through my work for the badger group that the chance to purchase a plot of land came up. I had been looking for some time. My experience of conservation work with the RSPB led to the dream of my own small-scale project. Parcels of land, private woods and fields were beginning to appear on the market, but I found myself up against rivals with serious money and schemes for making more. A wood, already compromised by its surroundings, could be used for rearing pheasants destined to be shot; it could be a source of timber; it could be a site for war games or a novel wedding reception or party venue, a campsite, an adventure playground with zip wires and tree houses, a proving ground for off-road vehicles . . .

During a routine visit to a big, long-established sett on private land my break came out of the blue. In passing, the widowed landowner revealed his intention to sell up the farmstead and downsize, and voiced his concerns regarding the badgers' future. He might, he announced, donate the discrete piece of land including and surrounding the sett to the National Trust. It was, he added, worth *x*-thousand pounds. 'I'll give you that for it,' I piped up. And that was pretty much that. The legal process dragged on due to pandemic restrictions but by the middle of summer I took possession of three and a half acres comprising a large field of permanent pasture, a young plantation surrounding a fenced compound with sheds, and, crucially, the clough.

2

Under the Skin

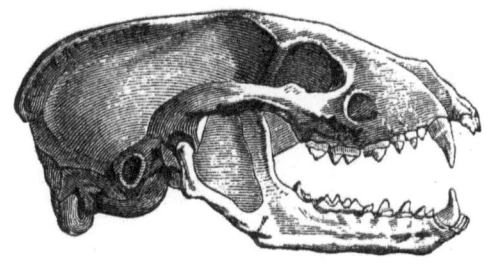

NOT ALL THE casualties are dead. I was given a crush cage and grasper. This type of carrying cage has a movable panel – the crusher – which, when deployed, holds a captive animal fast against the sides so that sedative, antibiotic or pain-killing injections can be administered. A grasper is a long metal pole with a noose on one end which can be tightened or released by manipulating mechanisms in the handle and shaft. More often than not, where badgers are concerned, the grasper ends up being used merely to prod with. This is because a badger has a wedge-shaped head and thick neck which enables it to slip the noose by the simple expedient of going into reverse. Further essential equipment for handling wild beasts in distress: a strong biting stick – a section of broom handle is ideal – and a supply of old towels and blankets. Injured animals can be (read: 'are usually') tricky to catch and handle, and it is advisable that more than one person turn out to deal with the situation. My dependable rapid response backup has come from Pam, long-serving chair, and her husband Philip, as good as an honorary member. On the occasions when an injured, as opposed to a dead, animal is reported, I will phone Pam and announce, 'We've got a live one.'

Most people will tell you, or each other, that badgers are aggressive and have a strong bite, but they have no firm grounds on which to base these claims. The second happens to be true. Historically, almost the only people who could boast of genuine close encounters with badgers were those, already biased against them, who persecuted and severely provoked them. If an injured badger can get away before humans try and intervene, it will. Most likely it will go to ground and die a lingering death. A crippled casualty unable to escape the well-meaning attentions of its would-be saviours will usually curl itself up, hiding its striped face. This is the opposite of a display of aggression. The badger may be trying to protect its vulnerable muzzle. The impact of a speeding car wheel on the muzzle is probably the cause of most more or less instantaneous deaths. Having developed the gruesome habit of inspecting the decomposing remains of erstwhile victims, I have found that the remarkable skull is, however, often intact.

The supposed aggressive nature of badgers is apocryphal but the power of their bite is legendary with good reason. In the earliest monograph on badgers, published in 1898, Alfred Pease relates how a gamekeeper engaged in extricating a badger from its burrow grasped a foot instead of the intended tail and was so tenaciously gripped in return that his hand had later to be amputated. At this distance, the present-day badger enthusiast can feel comfortably smug in the reflection that just deserts were starkly meted out on this occasion.

More than one unique feature of the skull is responsible for this fearful potential. The most striking singularity is the prominent ridge or sagittal crest, so named for the resemblance of its junction in plan view to an arrowhead (cf. Sagittarius). A fireman's helmet has a corresponding feature on top and, similarly, the crest may offer the burrowing mammal some protection from the occupational hazard of roof falls, but its primary evolved function is as a plane of anchorage for powerful jaw muscles. Secondly, a badger's jaws are both fused and locked, fused at the distal ends (like ours) and locked proximally where they are hinged with the upper skull.

An anthologised story I read in my youth – 'Meles Vulgaris' by Patrick Boyle – made a strong impression on me. It is an upsetting tale of badger-baiting in Ireland in which the progress of the despicable affair is related by the author to the animal's natural history and biology. The protection the sagittal crest, or interparietal ridge, gives to the top of the skull is probably exaggerated, however, when the tormented badger in the story survives a violent blow to the head from the blade of a mattock.

Once the insects have done their work on the flesh, the skull can be recovered for thorough cleaning (assuming a fox hasn't run off with it). Badger skulls are well worth preserving for their curiosity value. I recommend two courses of soaking each of two or three days, the first in dilute household bleach, the second in hydrogen peroxide solution. Hydrogen peroxide is widely and cheaply available from pharmacists, but I am often viewed with suspicion and only allowed one bottle at a time. It is, I believe, an ingredient in home-made bomb-making. The skull is such a labyrinthine grotto of hidden passageways that a few maggots survive even this harsh treatment, entrenched within deep air pockets and only emerging later, seemingly dazed. Look into the broad nasal cavity and marvel at the incomparable intricacy of the turbinals. These dense, highly convoluted structures of wafer-thin bone greatly increase the surface area of the inner membranes. Here, then, lies the secret of the badger's prodigious sense of smell. More than any other factor, this capacity is what makes the animal tick.

Most of the teeth – the single- and double-rooted ones – are likely to fall out during the cleaning process and will need to be superglued back in place. The finished skull should be labelled with details of the circumstances of the badger's demise since possession of their body parts is unlawful unless accidental or natural death can be authenticated.

Those of us who monitor setts may occasionally come across a skull in the vicinity of a sett. The explanation is likely to be, in the first instance, that a mortally wounded or sick animal instinctively went to ground, or remained below, in effect burying itself prior to its demise. It is believed that such corpses are walled up by the living occupants of the sett, presumably for hygienic reasons. At a later date the bones may be exhumed and cast out of the burrow along with other 'rubbish' (including used bedding). Generally speaking, these skulls will have at least a few teeth missing due to rough handling. Once, however, I was lucky enough to find a badger skull on a spoil heap which was caked in clayey soil, now dry and set hard, and all the teeth were firmly held in place by the substrate. The number of road casualties which survive long enough to make it back to their burrows then die underground constitutes an unknown quantity, but I have seen ample evidence to suggest this must sometimes happen.

At one time I supplied a friend with sheep skulls which I found on the moors. He liked to decorate them as artworks, and sometimes sold them for quite a lot of money. His cleaning method was different from that which I later adopted with badger skulls. A sheep skull is much larger and would require proportionately a lot more cleaning agent. When a skull was still almost fit for soup, he would soak it in a bucket of plain water in the greenhouse. The remnants of flesh macerated – became soft – and eventually fell away while a vile smell was generated. When he was satisfied, the bone was rinsed and scrubbed and dried out. Then he painted it white all over so that it became in effect a blank canvas before letting his creativity run riot with garish colours, decals, cheap bling and the like. He particularly liked to work with horned skulls. The finished articles had the flavour of tribal masks and were rather splendid.

A sheep skull, as befits a wholly dissimilar beast, is radically different from a badger skull. Nonetheless, the comparison is instructive. The eye sockets are framed by simple circles whereas the eyeballs of the piggy-eyed badger share space with bulky jaw muscles and are caged in by the zygomatic arches. On the crown of the skull, three crazy fractures meet up, again in the shape of an arrowhead, as if the sections of bone have been zipped together. The muzzle ends in long sharp points. Inside the nasal cavity the simple turbinals are scrolled like parchment. The jawbones are separate from the upper skull and from one another having in life been joined by cartilage and muscle, whereas a badger skull remains articulated in death. Ruminants lack upper incisors, the lower ones biting instead against a hard palate. There are no canines as such and little in the way of specialisation in the molars and premolars, reflecting a monotonous diet. A badger's diet, on the other hand, is very varied.

I have amassed a variety of skulls over time: sheep, cow, pig, red deer, roe deer, Chinese water deer, chamois (from friends with a skiing business), dog, cat, fox, badger, mink, hare, heron, porpoise (from a Cornish beach). The one which best bears comparison with the badger is the mink since both belong to the weasel family (Mustelidae), but whereas the mink is a typical mustelid, the badger is atypical. Their lifestyles are very different. Most mustelids have a long, slinky body plan and are designed to pursue smaller, agile vertebrate prey into tight corners. The badger, on the other hand, is stocky and powerful. Its modus operandi is to sniff out static or slowly moving food items, often subterranean, and, where necessary, dig them up with long claws. (The claws on the front feet are one of the things that strike you as remarkable when you handle your first dead badger. The forepaws are multi-purpose tools: picks and shovels for digging; rakes for collecting bedding material; combs for grooming; grappling irons for climbing; jemmies for prising open rolled-up hedgehogs; and weapons for slashing and scratching.)

The mink skull, in pieces, was given to me by a friend who had found the sorry drowned animal in a submerged trap. I cleaned the pieces up and fitted them back together. Anything smaller – a stoat or a weasel, say, with their needle-sharp canines – would require the patience and dexterity of a watchmaker. The vestiges or outline of a sagittal crest can be seen on the dorsal surface, but the skull is flat-topped since a high crest would be a liability in a predator which relies on the freedom provided by low head clearance. The mink's zygomatic arches are thin, curvaceous and fragile-looking; those of a badger are solid, rugged, robust. The mink has well-developed carnassial teeth for shearing through flesh and bone, more like those of a fox than a badger.

It isn't just me. When the opportunity arises to show off one of my badger skulls, the beholder is likely to respond with awe. This is partly because of the living animal's elusiveness, partly because what's beneath any organism's skin has a special fascination for us, and partly because a badger skull has the added wow factor of integrity.

Dealing with road casualties would bring countless opportunities for familiarising myself with carcasses and remains. But I also wanted to get to grips with the living creature, the fabled beast. The first time I loaded the cage and grasper into the back of my car, Pam and Philip weren't around. I wasn't responding to a casualty from an RTA or RTC (road traffic accident or collision – also shorthand for the victim) but to an out-of-place badger, a different category of customer but possibly with a degree of overlap. An out-of-place individual is one that turns up somewhere unexpected, often under, behind or even inside an outbuilding, usually discovered in a tight corner where it is difficult to reach. The reason for this behaviour may be an injury, perhaps from an RTA, or sickness. The commonest type is a victimised animal, usually a runt, which has been savaged by a dominant badger. Badgers can be very aggressive towards their own kind at stressful times. When the contest is evenly matched, wounds are usually inflicted on the neck and these are typically not serious, healing as battle scars. But when a subordinate is bullied, perhaps repeatedly, and turns its back on the aggressor, vicious wounds are sustained on the rump. Because of the badger's fossorial habits, these open wounds can become very dirty and infected. They are often lethal.

My first challenge, then, was to capture a badger which had entered an indoor sports facility and was last seen disappearing behind a vending machine in the stairwell. When I crawled into the dark space with a torch, I found the creature, just a small cub, curled up asleep in a makeshift bed of bubble-wrap sheets. Apparently, a gang of youths had come across the animal in the car park and their infectious excitement had driven it towards the sanctuary of the building. The main difficulty now proved to be getting the open cage into position so that the cub could be coaxed inside with the grasper in prod mode. Once I had it and had carried the cage out in the open, the cage seemed ludicrously roomy, more like a coop. When the time came the cub proved too small for the crusher to have any effect on its freedom of movement. Surrounded by spectators, the pathetic captive whimpered, it urinated and defecated through the mesh on to the floor. Yet it also snaffled up peanuts from a saucer I had slipped inside. At the vets the cub, a male, was discovered to have a congenital deformity, a hare lip, which prevented it from feeding normally and presumably accounted for its poor condition – it was drastically underweight – and plight. The vet recommended it be put down, which knocked the wind out of my sense of achievement at catching my first live badger. This was to become a depressingly familiar outcome.

Out-of-place badgers include those seen at large during the day when they ought to be underground. Again, there is always the suspicion that there must be something wrong with them. During a summer heatwave, I was called in regard to a badger which had been seen out in the open in the middle of the afternoon by a woman walking her dog. Pam and I went to investigate but, by the time we got there, the badger was nowhere to be seen. Eventually, however, we tracked it down through disturbances of the long grass to a vacant pig ark in an overgrown paddock where it was now curled up asleep in a dark corner. Playing a torch-beam over it, I could see that at least its coat was in good condition and, when questioned, the woman reported that its movements seemed normal. Since there was no obvious cause for concern, and the badger was safe for now, we chose not to disturb it but to leave it food, water and fresh straw for bedding.

The next day I found it sleeping just inside the entrance. I could have reached in and touched, even stroked it. The sleeper stirred a little as I replenished its provisions. I worried that it might overheat in this corrugated metal sweat-box. On the third morning it had gone. Since we had no clear idea where the badger had come from or where it was headed, it seemed best to let it work out for itself its next move, though I confess to being rather disappointed when I could no longer call myself its carer.

Capturing a cub is one thing, but an adult badger is a powerful animal weighing up to 15 kilograms and more, depending on the season. Badgers indulge in hyperphagy, that is, like the bears they resemble somewhat, they bulk up on autumn's abundance to see them through the winter. By the end of a good fruiting season they can be very heavy indeed. The largest corpse I ever handled was an exceptional male of 18.5 kilograms in mid-October, far too big to be weighed or carried in a supermarket 'bag for life', my standard receptacle. I fashioned a sort of harness out of bungee cords and weighed him with the spring balance that way. In the course of time I had his skull. Male badgers, or boars, outweigh females, sows, by one to two kilograms on average (see Appendix 2).

When Pam and I turned out in response to a particular report of an injured badger at the roadside, it was again mid-October, the heavy season. There was a trench resembling a fresh wheel rut in the grassed verge and the badger, a big fat specimen, was lodged in one end of it like a limpet in a crevice. The stricken animal had gouged out the trench itself with its forepaws in a desperate attempt to hide from the passing traffic. Now it was exhausted. There was an issue with its hindlegs which prevented it from crossing the narrow road back to its sett. After some debate, we decided to prise the badger out of its niche with a couple of biting sticks. It was a bit like trying to manoeuvre a sack of potatoes with outsized chopsticks. However, we were successful and, thanks to its exhausted state, we were able to manhandle it into the cage. Only once the badger was inside did it emit a low growl to let us know it hadn't enjoyed the experience. The captive almost filled the cage so that when the vet came to deliver the anti-inflammatory and sedative injections into its rump the crusher was redundant.

The casualty was a sow of 14.9 kilograms. X-rays showed that she had sustained a close transverse fracture to the left femur. This was relatively good news; the discovery of damage to the spine or pelvis would have been a death sentence. The fracture was hypothetically survivable, but to recover the use of the broken leg she would require three months' close confinement. With rest and luck, the break would heal and become callused, though the affected hindleg would be a little shorter.

The patient was named Bertha, and while housed at the local rescue centre she astounded everyone by climbing out of her stall in the night. The enclosure was fit for purpose, however, and she was unable to escape any further. There was life in the old girl yet and after a couple of days she was eating with gusto.

At the end of the week Bertha was relocated to Whitby for long-term rehabilitation, though she would be coming back to us once she was fully mobile. When she gave birth to three cubs in February while still in solitary confinement, we were delighted but not particularly surprised. True gestation lasts around two months, but badgers practise delayed implantation as a breeding strategy and she may have mated successfully as early as the previous spring. Our joy was short-lived, however, as first one cub then the remaining two all died for no obvious reason.

Bertha's return was delayed, first by the brief lives of her cubs and then by an outbreak of avian flu at her temporary home, and it was not until mid-April that the big day finally arrived, six months after her rescue. But without regular peer musking she had lost her clan identity. Sadly, she couldn't go back to her home sett. However, we had a vacant artificial sett ready for her which was constructed in tried and tested fashion by an experienced hired crew. To one of the two entrances we attached a release pen while the other was blocked with weld-mesh so that the sett was still ventilated. The idea was to keep her imprisoned and provisioned until she showed signs that she was settled. When releasing a captive badger into a new home, a useful ruse is to take some of its dung and soiled bedding – which I always request in advance – and smear it around the unfamiliar quarters. Recognising its own smell, the animal should readily enter the pen and the sett. This was happily the case with Bertha, who was looking none the worse for her traumatic recent past. She had recovered from her injury so completely that it took three people the best part of an hour to get her in the carrying cage at the other end.

Badgers are determined escapologists – and determined avoiders of human influence – and a couple of days later it was discovered that Bertha had dug her way out at the back of the sett. If her cubs had survived, would she have led them away or, in the role of mother, would she have settled? I have since come to accept, nonetheless, that all is not lost should this eventuality occur. A second rereleased sow did the same thing later in the year, but both badgers eventually returned, at least for a time, to the artificial homes provided for them. It is therefore a good idea to keep putting food and water out to encourage a badger to take up residence of its own accord and in its own time.

3

Underground

Soon after joining the badger group, I met Derek. He wasn't really a member, being from out of the area; he was more of a hanger-on. But his impact was great if sometimes unacknowledged, even resented in some quarters. He strayed into our group's territory because the area appealed to him for a number of reasons. He was interested in its particular industrial archaeology, he thought it more civilised than his home ground because there was markedly less badger persecution, and he was always on the lookout to extend his sphere of influence. Besides, there was little enough field activity going on and he was sucked in as though into a vacuum.

Derek is an ex-miner. It was his custom, like so many of his workmates, to chew tobacco underground to stave off thirst. The long-term result was that his teeth were ruined and the bulk of them fell out, and he can no longer bite into an apple. And the habit of doing without fluid all day stuck – while out badgering, which is what he does all the hours of daylight, almost every day – although he quit the tobacco. 'Do you want a swig, Derek?' I say, offering him a drink. 'I'm all right,' is the inevitable negative response.

He is interested, it seems, in all things subterranean, including badgers and all the places they inhabit, their natural setts and man-made substitutes: old mine workings, field drains, disused quarries, railway embankments, derelict cellars and ash tips; especially the latter. Badgers are an obsession with him, they seem to monopolise his sympathies. He is also a serious bottle-digger. Serious bottle-diggers go at ash tips, with or without permission, armed with spades and wheelbarrows, and excavate shafts so deep they need ladders to descend them. They know what treasures they are looking for and where to find them. They can date a tip or a find to a decade. One of Derek's chums has made – makes – a living from the practice and is, in his fifties, as happy as a man can be, and as fit as a fiddle. Derek, meanwhile – in his sixties – has arthritic knees which give him gyp over stiles and fences. I never knew anyone with such utter disregard for private land and prohibitive signs. Like a wild animal, he knows no boundaries and follows field edges and 'badger paths' wherever they take him. With two terriers in tow, he certainly ought to look suspicious but is seldom challenged. I suppose, what with his imposing size, lack of teeth and air of complacency, he looks a bit of a rough customer. Indeed he resembles nothing so much as 'the enemy'. He has familiarised himself on the ground with every feature and contour on maps of his area and ours. He knows every wood, path, field corner, shelter belt, piece of waste ground, brownfield site; every breach in municipal fencing; every sett and every ash tip. As for actual printed maps, he wears them out with use the way other people get through socks. By studying the maps and applying his accumulated field knowledge, he can predict where setts are likely to be.

Derek has a claim to fame, a literally accidental appearance on telly, on the local news. This happened in the third year of our companionship. While at the bottom of a shaft of his own digging the walls collapsed on him and he was buried alive. A friend tried to dig him out but struck him on the head with the spade while doing so. Derek was unconscious for a time and related afterwards how he had felt at peace for the first time in his life: he was sorry, he said, to 'come back'. The emergency services arrived in force but were uncertain as to how to proceed once Derek's bleeding head and upper body were clear since the ground was precarious and might give way again at any moment. Rescuers laid out planks and lay along them as though reaching out across quicksand. While the experts debated their next move, Derek freed an arm and, for want of anything better to do, began to dig himself out with a Victorian ceramic jam pot which had come to hand. When, eventually, he was pulled clear by a team of rescuers, he came out 'like a champagne cork'. He walked out of hospital the next day little the worse for his experience, certainly undaunted.

The terriers shared a branched lead and were dragged willy-nilly through bramble thickets. For one of the dogs Derek showed a strong affection, but he was forever cursing and clouting the other. Yet it was the latter who was 'keen', that let Derek know if a sett was occupied or not, while the beloved dog sat looking in another direction with no interest in the proceedings whatever. The keen, woolly-coated dog, Molly, seemed to like me, and perked up out of a sulk when she saw me coming, probably because I was sympathetic towards her and gave her special attention, but also undoubtedly because she could always smell badgers – dead ones – on my clothing. Any other dog encountered was snarled and yapped at, the terriers straining on the leash like a pint-sized Cerberus minus a head. Molly was the younger dog but it was she who died suddenly during my stint with the badger group.

Derek's speciality is setts which have been dug by baiters or other persecutors. It takes a practised eye to read the subtle signs and only his conviction will convince an inexperienced doubter who sees nothing amiss. Just as conscientious bottle-diggers backfill their excavations to repair the worst of the damage, canny badger-baiters hide the evidence of their dirty deeds. In order to obtain a badger, a terrier trained to corner and bark at one underground – but to avoid engaging it in outright combat – is inserted into a sett. The badger, although capable of tearing the little dog to pieces, attempts to escape by digging a way out, but turning its back on the dog gives the latter the advantage. The terrier wears a locater collar and, above ground, a receiver picks up a signal which tells the baiters where to dig down. My understanding is that when a badger has been successfully extracted, it is set upon by more than one dog – lurchers have powerful jaws and are too quick for a badger – the whole sordid business being filmed with a phone. The hole is typically backfilled, but it is possible to see evidence of square cuts in the ground and severed tree roots if one looks carefully.

Derek organises sett protections where sheets of weld-mesh and thick rubber belting are laid beneath the topsoil, as well as overseeing the construction of artificial setts. Badger groups raise funds by various means to pay for such projects.

We can monitor a badger sett by day and assess the level of activity based on the freshness and extent of recognised signs. We can examine the corpse of a road casualty and make assumptions about where it came from and was headed, and about the purpose of its rudely interrupted journey. And we can watch badgers emerge from their burrows at dusk and, if we are lucky, see how they behave. Yet they remain essentially mysterious. We can't see inside their underground homes – not least because their burrows often begin with a dogleg – and can't follow them in the dark through thick vegetation. We can barely make sense of their actions even when they are briefly in view. Sometimes the environs of a sett will reveal frustratingly few clues, and Derek for one will start jumping to conclusions about interference by the hunt, or a local farmer, or, rather vaguely, 'scum'. Our evidence will always be circumstantial; we can never know for sure what or if anything has occurred. Badger groups have got their work cut out.

As for attempts to prosecute badger persecutors, they are often undermined by the convenient excuse, 'I was after foxes, your honour.'

The home ranges and operations of badgers and foxes very often overlap. Where two large carnivores occupy the same ecosystem their habits and adaptations differ to avoid direct competition for the same resources; nevertheless they are bound to come into conflict, and then fur may fly. Being much heavier, badgers generally intimidate foxes and the latter are quick to take evasive action.

Compared with a badger, a fox has a range of acute senses, is long-legged and has an extravagant tail. Compared with a badger, a fox has an expressive face. Foxes have vertical oval pupils like cats and unlike larger canids. The ears are very mobile and can be moved independently; I obtained evidence of this fact in a series of trail camera stills. A fox is countershaded in the usual way of mammals, being lighter underneath, whereas a badger's underside, though more sparsely covered, is of darker hair. Where the two animals visit the same food source their behaviours are different. A local 'mad cat woman' who also puts food out for wild creatures buys raw chicken pieces in bulk for her nocturnal visitors. The foxes are out first, and on entering her back garden they pick up their prizes and carry them away. On the late shift, however, the badgers eat their meat where they find it. Although the situation is artificial, these actions have their natural analogues. While a badger has evolved the capacity to make the most of a windfall at one sitting, a fox's response to abundance is to cache for later what it can't eat now.

Badgers inhabit their systems of burrows year round but foxes spend less time underground than is generally supposed. During the summer, when the undergrowth is thick enough to hide them, they prefer to lie up above ground. Badgers are industrious earthmovers and gather bedding to make underground nests. Vixens often choose to raise their cubs in vacant badger setts, taking advantage of the labour of others. While raising their cubs, both vixen and dog fox bring back food to the den, something which parent badgers have rarely been observed to do. A badger sett occupied by foxes isn't hard to distinguish. For one thing, there's the acrid smell of fox. For another, the ground is likely to be littered with food remains, especially feathers. When the surrounding vegetation is flattened, it is a sure sign that the cubs have reached the exploratory/playful stage.

The owners of a private wood were concerned when they discovered what they took to be small badger cubs above ground one freezing cold April morning. There were four tiny, dark creatures on the scant woodland floor outside a hole and they were behaving rather strangely. 'Those aren't badger cubs,' I said, 'they're fox cubs.' Very young cubs are not especially foxy. Their coats are woolly and chocolate brown rather than russet; they have short faces, blue eyes and puppy-dog tails. At this stage a fox cub might be compared to a larva, dissimilar to the imago form, while an equivalent badger cub is comparable to a nymph or, in the parlance of popular culture, a 'mini-me'. These fox 'larvae' were squabbling with pathetic ferocity over the meagre remains of a dead crow, which they were probably incapable of assimilating, and were quite oblivious of their surroundings.

After a while we were joined by Derek, Pam and Adam, Pam's right-hand man in the group. We all stood about ridiculously tall while these strange miniature beings from the underworld enacted a desperate-seeming drama at our feet.

Something was very wrong; these cubs were too young to be above ground yet, and even when they were of an age they would only emerge during the day to bask in warm sunshine. Young mammals which are born in a burrow undergo a sort of second birth – from the womb of the earth – when they are ready to venture above ground for the first time. In the case of foxes, this occurs at about four weeks. I didn't think these cubs were yet that old. The only conclusion to be drawn was that the vixen – perhaps both parents – had come to grief and that these cubs were orphans. On such a bitter day they were sure to die.

Blue-eyed cubs are short-sighted as well as unwary. As if claiming the pick of a litter of puppies, Derek, who had left the terriers in his van, was the first to kneel down and lift up one of the cubs. It snarled like a kitten pretending to be a tiger. 'You little beauty,' he crowed, squeezing it in his fingers like a fistful of clay. When I handled one, it tried to bite but its minuscule teeth seemed incapable of breaking human skin.

After I had consulted over the phone with a fox specialist, we began to bundle the cubs together inside Pam's spare fleece and stuff them into a hastily emptied rucksack. Then I told everyone to be quiet – I could hear another cub still below ground. Derek reached in at full length and extracted cub number five.

Pam and I took them to the local wildlife rescue centre where they were placed in an incubator. Some time later they were transferred to a larger enclosure and presented with a bowl of tinned puppy food and another of puppy milk, which is low in lactose. While the rest mostly fought over the former, one cub stood in the bowl of milk messily lapping away. Happily, as the specialist had explained, fox cubs are able to lap milk from an early age. After a while they were all sodden and had to be returned to the incubator to dry out. Later the same day they were moved to a fox sanctuary in another county. Successive bulletins on their progress over the following months were all good. When the time was right, they would be released back into the wild.

In those early days Derek often drove us all – the hard-core of the group – around in his trusty Transit van, the faulty side loading door of which would slide wide open as the vehicle ground uphill. He didn't drive much faster on the flat since he was forever, while on the road, peering out of the windows on the lookout for badger paths where they intersected with the tarmac – those well-trodden paths which when followed led to the setts he knew of, or perhaps exciting new discoveries. The traffic built up behind us but Derek paid it no heed.

One time I was asked if I might rescue an adult fox caught in a trap. Cage traps as well as snares set for foxes (but capable of holding badgers) are perfectly legal if not logical since if one sentient, demonstrably intelligent large mammal should be legally protected from harm then why not another? This is called philosophy and it is an example of human beings at their best. I was about to break the law on philosophical as well as compassionate grounds.

For moral support and possible assistance, I asked Derek to come along, and we set out in his van. The cage trap was in the middle of a field next door to a ramshackle chicken run. It was big enough and strong enough to hold a bear. The fox, a lovely vixen, seemed to be patiently waiting for someone to come and open the cage, but it would be someone with a loaded gun. Fortunately, that person was nowhere to be seen. When I squatted down and fumbled with the catch to the spring-loaded door, the fox leapt up the back of the cage and tried to get through the roof. Once I had the heavy door up a fraction, I signalled to Derek to bring his dogs round and flush her out. She looked at the dogs, then distrustfully at me. Then she saw the gap under the door. She decided to take the risk, squeezed under and was away up the field, a red streak of freedom.

Derek boasted that we would come back and trash the trap so I held him to his word, but privately I had my doubts. We returned after dark. Derek had brought a sledgehammer as promised. He was going to reduce the offending article to atoms. I stayed in the van with the dogs while my partner in crime slipped out with the atomiser under his jacket. He was gone some time. The dogs whined for their absent master. An occasional car, headlights on full, crawled past our parking place on the unmetalled track. Otherwise all was still and quiet. After ten painfully slow minutes Derek was back looking even more sheepish than when he'd set out. He admitted that he 'couldn't put a dint' in the sturdy frame of the trap, which had been reset with half a chicken as bait – and had only succeeded in springing the door shut. At least, I consoled him, no more foxes would be caught that night.

4

The Compound

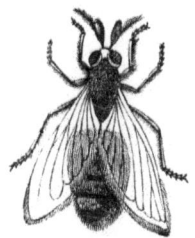

IN DUE COURSE Derek's Transit-van-cum-minibus gave up the ghost and was replaced by a two-seater van. We took to going out badgering in two vehicles, Derek's and mine, but there were further occasions when it was just me and Derek and I occupied the passenger seat in his van. This was like sitting inside a skip since there was so much rubbish on the floor. When I opened the door crisp packets and paper bags would be whisked out on to the verge or pavement and I would be forced to round them up once more. The dogs were in the back, of course. Mostly they were quiet, but as soon as Derek carelessly tossed morsels from his lunch over his shoulder, they would fight savagely and Derek's yelling only added to the furore.

Derek made a point of not washing his hands before eating. This fact impressed me early on when, during one of our first outings together, he had pushed his fingers into the contents of a badger's dung-pit, lifted some of the squidgy mess to his nose and declared that the smell was similar to that of cow manure. Later, when we sat down to our packed lunches on a low wall, I offered him a wet wipe, but he declined.

If there were three or four of us in the party, Pam would be my passenger, ready to lend assistance if I was called away to a casualty. One time, it was just the three of us and, out in the field, we had split up to cover a wider area. From somewhere in the distance, I became aware of a sound that was spoiling the tranquillity of the day and the location. Some idiot had apparently lost control of his dog and was shouting its name over and over. 'Molly' – the name was common enough. It was some time before I realised that the infuriating bellowing voice was Derek's. I homed in on it. Molly had gone missing and, although she was the object of his casual abuse, Derek had completely lost his composure. I tried to get some sense out of him. He seemed to think she had gone down a fox earth at the bottom of a steep rockface. I climbed down using tough heather stems as handholds. It was spring and the earth was active. There would be cubs inside. There was a dead mole outside one of three entrances. Foxes, like other predators, kill moles instinctively when they come across them, but finding them distasteful they become playthings for the cubs. All was quiet. I widened my search while Derek went on showing us up with his distraught shouting. Meanwhile, Pam had returned from checking the van where she had found no sign of the errant terrier. Eventually I ran into a couple of women who had come across a lost dog which was, at that moment, secreted in their car. When I claimed it, they demanded a description, which I was able to provide to their satisfaction. When Molly saw me, she wagged her stumpy tail sheepishly, sensing that she was in trouble. I picked her up and thanked the women. As Derek came marching up, I shielded the dog, knowing she was going to be whacked. At last, Derek calmed down and was pathetically grateful that I had kept a level head.

From that day on, Molly was seldom allowed off the lead. I sensed too that there had been a shift in mine and Derek's relationship. No longer was I his pupil; I had now to accept the responsibility of being his minder and advisor. I felt a bit like the straight guy in a vaudeville double act. Whenever he repeated one of his tall tales, I reminded him that we had heard that one several times before. A black fox; mysterious lights in the night sky; a crypt beneath the foundations of a demolished hospital; a gannet blown off course in his garden (I liked to joke that it had been on his nuts).

It all came about rather quickly.

An old friend retired a year later than I did. I thought we might see a lot of each other, but it didn't work out that way. But we did spend a month as travelling companions in India.

I should say that I *accompanied* Laurence since he was totally in charge of the arrangements, the transactions and the itinerary. He was interested in the people, I in the animals. He, an experienced globetrotter, was completely at ease, whereas I was wound up tight by all the attention we attracted, wholly unwanted on my part. We were expected to pose for selfies and be amenable to being guided on a whim, and to be susceptible to buying every souvenir, to bestowing alms on every beggar. I was horrified by the extent to which human beings could overrun and despoil a country – such a vast country at that. The interminable temples were at least sanctuaries for wildlife used to the human throng, and I told myself that the insufferable heat and the appalling smell, the harassment and the squalor, were all worth it for the monkeys, the colourful birds, the gaudy butterflies and the street dogs. Nevertheless, I was glad to get back to peaceful old England. But when I saw through fresh eyes the way in which our own countryside was treated as a rubbish dump despite all our advantages, I felt deeply ashamed. At the very least, I would spend the rest of my active life picking up litter. More than ever, I was determined to be useful, to try and make some sort of difference.

It was Laurence who planted another seed, when he happened to mention that his long-time best friends fed badgers in their garden every night. I couldn't believe he'd never thought to tell me this before. At my insistence, he got me invited. I took a bottle of wine and as darkness began to fall I was seated in an armchair facing the French windows. On the other side of the glass there was a wide step above the back patio. A generous heap of peanuts had been placed on the step. I plugged my hosts, the friends of the friend, with questions while we waited. They told me that they kept their wild guests secret from their neighbours. At the back of the bordered lawn ran a low wall with a purpose-built gap towards one end. Beyond the wall was, apparently, a field of pasture which had been earmarked for development; but this plot, too, had been part of a field once. And before that?

Long after dark, the automatic light came on outside. Something had entered the garden through the gap and was now scuttling around the perimeter, grey and dark, like a giant rat. Then it came right up on to the step and into clear view, a badger with a black-and-white striped face. I leaned forward in my seat and stared in wonder. Some minutes later a second badger entered the arena and the first one immediately took fright, abandoning the nuts, completing a circuit of the garden and exiting the way it had come. The boss had clearly arrived. This was confirmed when a third badger showed up and was prevented from accessing the bulk of the nuts by the boss's broad rear, which was swung in the other's face like a ram. (One thing I learnt in retrospect from that night was to present food to badgers in separate piles.) When the show was over and the players had departed into the mysterious night, I thanked my hosts and declared that it had been one of the best evenings I had ever spent, and meant it.

It is a measure of how much progress I made that, five years later, my hostess of that night invited me to give a talk and slideshow on badgers to her women's group.

In the meantime, badgering in its various forms had become an all-consuming way of life, a daily experience and pastime. I was fast-tracked by the hard-core colleagues I spent so much time with, especially Derek and Pam. I read numerous books on badgers – as well as foxes. I looked at the world around me from a different perspective, in terms of its connectedness for badgers rather than people. When my dreams weren't haunted by the recurrent revelation that I hadn't retired from teaching after all, I dreamt about badgers.

In conjunction with my growing obsession, I lost interest in so many other things which had once seemed to matter: sport, music, literature, cinema . . . More so than ever, I experienced difficulty in finding common ground with others, even my old friends. I had gone underground. I knew that I had changed radically and I hoped that it was a good, healthy thing rather than a sign of incipient dementia.

Derek took a great interest in my records of casualties, phoning almost every other day for the latest news. Each piece of information was another clue for him to follow up, and might lead to the discovery of a sett which had as yet eluded him. Of particular interest to me were the sex ratio of casualties, which times of the year yielded most accidents, and the weight of the victims in relation to the date of death. All these things, which relate to seasonal behaviour, had been studied before at a national level, and Pam referred me to the 'Badger Bible', Ernest Neal's 1986 monograph *The Natural History of Badgers*.

Once I could say that my data was comprehensive, that fact, combined with the group's sett records compiled by Pam and Adam and based on their and Derek's long-running field-work, might enable me to make a viable estimate of how many badgers there are in the region. It seemed to me that somebody ought to know the population size. Also, from a local perspective, mapping the casualties would give an idea of relative badger distribution. In the new year I handed my GPS data for the previous twelve months to a friend with GIS (geographic information system) skills who created a choropleth, a colour-coded map of the borough divided into its twenty-plus political wards. The stronger the colour – the bloodier in fact – the greater the incidence of roadkill and, perhaps, the greater the density of badgers. (There are flaws to these assumptions, however: the wards are not units of standard area, and neither is the road network or volume of traffic evenly distributed throughout. I know for a fact that one particular location with a very high badger population yields few casualties because it is relatively rural. Nevertheless, a regional choropleth can provide a useful overview.)

As Neal led me to expect, I was busiest in spring and autumn while winter was relatively quiet. Late February is when things really kick off, which coincides with the time when most mature sows give birth and become receptive to the males' sexual advances once more. Since boars may have to travel further afield to satisfy the urge to mate, they are more likely to succumb to road traffic.

I always responded promptly to each report as it came in. My habit was to reach out of bed and switch on my phone before rising to check for messages or await incoming calls. These might be from the council, police or general public. The quicker I could get to a casualty the more likely it would be in good condition and the more I could learn from it (this is one of the list of reasons for responding promptly – see Appendix 3). This often meant postponing breakfast and joining the morning rush hour, something I wouldn't have bargained on when I first retired. If the scene of an accident was a busy road then I slipped on a fluorescent bib bearing the group's logo. I was now identifiable as the Badger Man.

Not all the corpses I examined were fresh. Some had been

overlooked in the first instance and only noticed later by a chance-comer on foot, perhaps due to the foul smell. It wasn't easy to back-date these victims but I thought perhaps if I documented the progress of decay on a day-to-day basis then I might gain some useful insights. The rate of decomposition is heavily influenced by time of year and the scavengers involved. In order to observe gradual and predictable effects I needed to restrict the process to insects and invisible agents so I carried a good-sized fresh specimen, a twelve-kilogram boar, up to my compound, which is surrounded by a six-foot fence of close mesh. By placing the carcass beneath a low, leafy tree, I hoped to hide it from beady-eyed corvids. It was the end of July so I was expecting rapid results.

I laid the badger on its side so that I could see its underside. On that first day there was no taint of corruption to my nose, though the corpse was attracting flies immediately. To my mind, the smell of a badger is inoffensive, a sharp mustiness with hints of goat, wet dog and second-hand bookshop.

By the second day the corpse had attracted a host of admirers: clustered thickly on its throat like living jewellery, crawling over the paler abdomen and rear-end, sipping the moisture from its protruding tongue; one fat fly came out of the slightly open mouth. The flies were of three types: buzzing blow flies, orange dung flies in mating pairs and a few small, nondescript flies. The blow flies could be further subdivided into metallic blue- and greenbottles and larger, duller flies of otherwise similar appearance. The corpse's abdomen was already noticeably swollen, its pale skin livid in patches, and flies' eggs were visible through the sparse hairs of the belly like a dusting of sulphur powder. There was a whiff of putrefaction at close range. While the females jostled for position on the corpse, the male blow flies could be found on the creeping thistle flowers in my field along with the butterflies and bees and other pollinators, for the male flies have no taste for rotting flesh (male mosquitos also sip nectar and leave the blood-sucking to the females).

On the fourth day, the whole underside of the badger was thickly coated with flies' eggs as though daubed with creamy-yellow paint. Most of the egg-layers had departed now, perhaps they had died after fulfilling their purpose. The smell was still tolerable at close range if not exactly pleasant.

The common bluebottle goes by the charming scientific name *Calliphora vomitoria*. Its equally charming offspring, once hatched, secrete enzymes on to dead flesh which cause it to liquefy. They are equipped with special breathing apparatus to prevent them from drowning in a soup of their own making. On day six these maggots had migrated beneath the carcass where I could just detect their wriggling movements. Hardly any adult flies were left now and those that remained seemed stupefied, as though drunk. They had been joined by a solitary black-and-red sexton, or burying, beetle – not that even a swarm of them could bury a badger: I once lifted up a sheep carcass to reveal hundreds of burying beetle larvae (normally, a single mating pair of beetles inter something more manageable, such as a mouse or fallen sparrow, as food for their grubs). The badger's face had become fuzzy-looking and was losing its contrasts. For the first time, on arrival, I could smell the rotting corpse from outside the compound.

By day eight there was a constant seething of maggots like a pan of rice kept bubbling and the badger's hair had slipped (come loose) from where they roiled. Although the maggots had grown it was difficult to say what they had assimilated since the exterior of the corpse, though hairless, remained intact. Two days later the maggots – thousands of them – were all over the body like a living fleece and the mouth was stuffed with them. The dead body was starting to lose its form. The smell was particularly strong on day eleven, which perhaps wasn't surprising since a gaping cavity had appeared in the corpse's underside through which the gases of decomposition could escape. On day twelve most of the maggots had retreated inside the cavity, leaving behind a bald skin which was blackened and papery as if it had been burnt. The corpse lay in a muddy puddle of gloop and the stench was appalling.

Sometimes, once I get the smell of death in my nostrils, I can't seem to shake it, I go on smelling it; it seems to be on my fingers and everything I touch is contaminated. Soap won't shift it, but that's because the smell is only a memory and not really present.

There was more significant progress on day thirteen. Most of the maggots were no longer visible; presumably they had gone underground to pupate. The skull could now be seen and would soon be ready for me to 'harvest'. After two weeks the badger was a very sorry sight indeed. It still had legs, and parchment-like skin covered its ribs, but it was deflated and ragged and the colour of earth. Around this time a carcass elsewhere in a similar state had attracted several large rove beetles – *Creophilus* – banded black and grey: the scientific name translates as 'lover of flesh' but they are really drawn to dung and carrion to prey on other insects.

After three weeks the smell from the badger in the compound was no longer apparent, at least from a respectful distance. The skin that was left had retracted and the remains were shrunken, and *something* had rummaged among the bones and moved some of them a short distance away. Among the scattered spillikins was the baculum (penis bone), which I'm not often lucky enough to find. Small black flies, little bigger than midges, had now settled on the leftovers.

I took the baculum away and cleaned it up. It was nearly 8 centimetres long, like a fat bent needle with an eye in the spatulate end and a heavy point at the other; a shallow groove ran along its length on the convex side. The purpose of the baculum in mustelids and some other mammals (seals have spectacular bacula) is to help sustain bouts of long-duration mating – up to an hour and a half in badgers. The baculum 'floats' free, independent of the rest of the skeleton, and is easily lost during the natural process of disassembly. At one time they were worn as tie-pins as part of a posing badger persecutor's traditional (Scots) costume: the *pièce de résistance* was the poor creature's mask, which formed the flap for the sporran.

At the end of four weeks the badger was reduced to a wrung-out rag lying in a dark stain at the foot of the tree. It had little more to give. What tufts of hair remained were bleached and rotten, and the cracked skin came away from the skull like a dry husk from a nut. The once sturdy skeleton had collapsed into a jumble of brittle-looking sticks. Only the specialised skull, which I would clean up and display (or give away), retained any dignity. It would serve as a monument to the living animal. To a large extent a dead body is transformed into a swarm of flies, and the flies disperse and seek out the next corpse, reanimating it.

The compound encloses a 25-yard-square plot within the strip of plantation east of my field. The trees in and around the enclosure are small varieties: hazel, rowan, field maple, guelder rose, hawthorn and apple. There are lots of the latter and in a good year there will be hundreds of apples. Many are scrumped by the free-roaming sheep but the ones that fall inside the compound I leave for the fieldfares and blackbirds. Mistle-thrushes feast on the rowan berries. I also hang my bird feeders in the trees of the compound from late autumn until the following spring. A flock of jackdaws descends daily on the peanuts and fat balls, but they leave the sunflower hearts for the tits and finches. This artificial set-up creates a good opportunity to compare the feeding strategies of some of our familiar small birds. Coal tits, for example, expend a lot of energy toing and froing – they are scatter hoarders – while greenfinches and goldfinches hog the perches and consume as many seeds as they can before they are displaced or disturbed. Chaffinches mostly feed on spillage on the ground, as do bramblings, a small flock of which hung around one winter. Robins visit the feeders but, like chaffinches, are rather in-expert at using them. All these small birds must beware the sparrowhawk which is usually lurking somewhere nearby.

I had further plans for this fenced area. When short-term accommodation couldn't be found for an injured badger cub – the rescue centre was full – I thought one of the sheds in my compound would do. When we came upon the cub it was unable to get away and flapped on the spot like a fish out of water. The vet's diagnosis was 'dislocated carpus left fore'. The pedant in me thought, if it was left hind it would have been dislocated *tarsus*.

Things seemed to go well at first. The cub, a male, slept soundly in a dark corner of the shed among the pile of hay provided for him and during the night had sampled some of the food options on offer. And he had used the open carrying cage lined with an old blanket as a latrine. However, there were signs of restlessness and attempts to get out. He had scratched up the vinyl floor covering in the corners, reducing it to shreds, and had tested the barricade I had created inside the door.

Two capable volunteers from the rescue centre arrived next day to administer the cub's daily anti-inflammatory injection. We noticed that the cub was moving about more freely now, but it had been decided to relocate him to Whitby the next day where there was a solitary cub in residence that needed a companion.

When I entered the compound on the second morning, psyched up for the long road trip, I saw straight away that there was a gaping hole in the bottom of the shed door.

Inside, the barricade was down and none of the corners of the shed revealed what I was looking for. All the food had been eaten. Clearly the cub was loose but I still expected to find him somewhere in the fenced enclosure. I started with a circuit of the perimeter but could find no evidence of digging or any gap he could have squeezed through. Potentially there was no shortage of hiding places but I scoured every square yard of ground – more than once. In the end I had to consider the unlikely seeming scenario that, pumped up on drugs and adrenalin, the cub, so recently immobile, had effected a miraculous escape by scaling the high fence. I know that the skull of a cub is a work in progress, lacking as yet the robust features which distinguish the adult version, yet it is apparently resilient enough to enable a youngster to chomp its way through a wooden door. When another cub, a dehydrated female, was rescued two weeks later, I didn't delay in taking it to Whitby but set off straight away.

My next lodgers were also mustelids. I was entrusted with two rescued weasel kits, one after the other, for 'soft release' once they were old enough to tear day-old chicks to pieces for themselves. Soft release entails providing increasing amounts of freedom gradually and decreasing amounts of food. 'Hard release' is when a rescued animal is given its freedom back in an instant by the expedient of opening a carrying cage door. (I released a rehabilitated kestrel on my land in this manner, which needed no encouragement to launch itself into space, sailing right across the valley and out of sight.)

I kept the weasels, one at a time, in a hutch in the shed with the hole in the bottom of the door. I thought the hole might come in useful. By the time I moved on to the second weasel after a gap of a couple of weeks I had perfected the technique. Keep the creature cooped up in the hutch for a day or two to get it used to its new base, then leave the gap in the mesh front unblocked and let the weasel come out and explore the shed in its own time. When it is ready for the next step the tyro can begin to venture out into the big wide world beyond the shed. It can always return to the shed and the inner sanctum of the hutch – where there will be food waiting – if it feels the need.

I filled the hutch with twigs and cardboard tubes to enhance the weasel's exploratory development. As the smallest of carnivores, a weasel has the instinct to cache its kills promptly, so when a fresh chick was introduced it would be immediately taken into hiding. Or that was the plan. Inevitably, the chick would become lodged in the fork of a twig and the weasel would tug and tug and turn somersaults until, more by luck than judgement, the prize came free.

Both weasels seemed to enjoy my twice-daily visits to the shed. They came to associate me with food; they no longer chittered with annoyance or emitted high-pitched squeaks of alarm – like steel on steel – as they had done during the traumatic experience of moving them here from the rescue centre. Now they responded to company by showing off. These delightful small mustelids seemed to assume that all the items in the shed – the various tools, boxes and sacks of miscellaneous articles, and offcuts of wood – were put there expressly for their amusement and edification.

As with many small mammals, our 24-hour days are too long and exhausting for weasels so they divide them up into shorter periods of activity and rest. This makes them both nocturnal and diurnal. When awake, a weasel is like an adolescent on speed: hyper-alert, twitchy, nervous and paranoid. Sometimes when I opened the shed door and stepped inside, all would be quiet. At other times the weasel would be eager to play. The second weasel, even smaller than the first and probably female, was especially playful. Many of her lively antics involved whirling around. She would have mad sessions inside a ceramic jam pot (from a bottle tip), which was laid horizontal, doing impressions of a bundle of laundry inside a washing-machine; or in a dog bowl which held a few loose screws that gave added pleasure. A scrunched-up ball of newspaper had served as a plug for the gap in the mesh of the hutch. It now lay on one side. The unfurling paper ball was treated as if it were a dull-witted adversary, ambushed and, like all her other playthings, made fully aware of the sharpness of her teeth. Where two bulging plastic sacks leaned together and formed a slippery hollow, she flung herself into the hidden space and squirmed about while uttering a low chuckle.

The only flaw in my method was that as a weasel gained in confidence and, by stages, moved out, a rat would move in and, in its turn, expect to be fed on day-old chicks.

5

Borderlands

DEREK TOOK ME into zones where I wouldn't have ventured on my own, either dared or thought to. Not just brownfield sites where we sometimes stumbled on derelict cannabis-growing operations or homeless souls sleeping rough, but notorious housing estates where only the delusional – or visionary – would seek for badgers.

A typical feature of these sprawling, disconcerting estates is the unofficial dump in plain view, as brazen as a farmer's multi-generational junk heap. Unwanted furniture, white goods, bathroom and kitchen fittings, children's playthings in garish plastic, car parts, outmoded gadgets; all pile up on what may have been intended by town planners as communal green space. The scale of these tips quickly grows beyond the control (and interest) of over-stretched local authorities, and they become, in a sense, invisible and the behaviour normalised. At one time such multifarious heaps might have provided raw materials for local children's imaginative play, but there is little evidence of this now; most of the items on these modern middens have already exhausted their potential.

Hard-core wildlife enthusiasts may, among other things, be accidental urban explorers and become well acquainted with the landscape of neglect.

Lawless estates are firmly established in the consciousness of the general public, in modern mythology and popular culture. Perhaps the most powerful film to exploit our fear of such places and their antisocial inhabitants, whether imaginary or not, is *Eden Lake* (2008), a sort of parochial English-based *Deliverance* (the latter film is the touchstone for any on-screen adventure where townspeople find themselves out of their depth in an out-of-town setting). A handsome well-to-do young couple take advantage of the brownfield site of the film's title, a flooded quarry in an abandoned park scheduled for redevelopment as a gated community, accessed through a gap in municipal fencing. Here they plan to spend a romantic weekend camping, swimming and sunbathing in what appears to be a private and peaceful setting, but are disturbed by the noise and disruption from a gang of surly youths who soon turn up. The man is presumptuous enough to face up to them. Presently he will be tortured to death. Then the plot hinges on the resourcefulness of the heroine and her desperate attempts to get away with her life. Eventually she turns to the local residents for help, not realising that she has fallen into the hands of the parents of the youths, two of whom she has killed.

At some of the dodgy locations we stray into, looking for unlikely evidence of badgers in the urban badlands, I find myself exclaiming to Derek, who has also seen the film, 'Oh, my God! It's like *Eden Lake*.' In one of many hard-to-watch scenes in the film, the gang of youths are seen violently kicking a cage trap. Inside is a badger. Badgers make relatively few appearances in popular culture, at least for adults, and then they are typically portrayed as victims.

Our wildlife crime officer had arranged for Derek and me to appear on local radio. It was early spring so the purpose was ostensibly to draw the attention of motorists to height-ened badger activity at this time. Patrick, the talk show presenter, agreed to meet us in a pub car park on his way into work. He had brought his portable recording equipment along with him and, serendipitously, had spotted a dead badger just down the road. I wandered off to investigate while Derek regaled Patrick with badger lore. The casualty was a battered specimen lying in the gutter, a parcel of dark oxidised meat bundled in a dirty wrapper, just about identifiable as a young male. It was just as well that Patrick's audience was spared the sight. When it was my turn to speak, I talked statis-tics and behaviour patterns. Patrick conducted the interview with respectful interest. I thought it had gone well, not that I got to hear any of the resulting broadcast that was aired later the same day. Derek, however, had tuned in and was 'appalled'. The subject, he claimed, had been reduced to a joke by the addition of incongruous music and crass rhymes. And we came across as 'nerds'. He assured me that I would be 'aston-ished' at how bad it was. But I suppose we *were* nerds. We talked about badgers the way other people talk about football or cars.

The interview coincided with confirmation of the first case of coronavirus in the borough.

When the virus struck I was living in rented accom-modation in a different part of the borough while seeking, long-term, a new address in the valley I called home. My temporary lodging was in a tiny village surrounded by pasto-ral farmland. In contrast to my previous setting there was little in the way of surface water here, the moisture being locked up in the greenery instead. The centre of the village – church, shop, pub and primary school – was organised around a narrow through road where it turned a tight corner, and this unfortunate circumstance, together with the volume of through traffic, undermined the settlement's potential as a desirable location. At least twice a day there would be grid-lock as careless on-street parking – or the occasional funeral – blocked the passage of a double-decker bus, bin wagon or removal van. And the daily exodus from school was chaotic. Only during lockdown, with the closure of the school and the cessation of public transport and casual traffic, did the village live up to the favourable impression a transient outsider might otherwise have formed. Only now did its narrow ways feel like country lanes.

My humble abode was also situated on this troublesome corner, but diagonally opposite I had the advantage of the church's largely untended graveyard. Here I fed the birds in winter and here I would sit on a bench with a book on sunny days as if enjoying the privileges of a large, shambolic garden. In summer the old stones were draped and smoth-ered in bindweed, a decorative plant of neglected sites which gets a lot of bad press as an invasive weed. A pair of tawny owls roosted by day among the dense foliage of the evergreens, occasionally emitting diurnal contact calls: 'Where are you?' 'Over here.' There were squirrels and songbirds year round. The churchyard proved to be a godsend during lockdown.

I was looking for peace; I usually am. The village I had left behind was booming. Ever more wine bars, bistros, beauty salons and gift shops were opening to meet its pretensions. The pandemic came at a good time for me, and my usual lifestyle was pretty much unaffected (a friend said I was a trailblazer for social distancing). Apart from the fact that there were no more casualties to attend to. If I had fantasised that, in my role as the Badger Man, I could be another key worker given heroic status, I was to be disenchanted. I now had time on my hands. Not having access to the internet, and the bookshops being closed, I fell back on my own shelves and picked up books I hadn't got around to reading. The weather helped – it was the balmiest of springs. For the lucky ones, having a private garden made all the difference.

I still received a few calls about badgers. Some people take exception to wild creatures digging in their manicured lawns. During dry spells there is an increased likelihood of badgers grubbing for subterranean insect larvae: leatherjackets (crane-fly larvae), cutworms (the caterpillars of noctuid moths) and chafer grubs. One of my regular pro-badger callers told me about neighbours of hers who objected to badgers digging up their lawn and had invested in a camera trap to find out where the miscreants were getting in with a view to blocking their point of entry. As well as capturing footage of the badgers, the camera picked up the unexpected incursion of two men during the small hours of the night who proceeded to help themselves to the contents of the garden shed. Since the photographic evidence led to the apprehension of the burglars by the police and the subsequent return of the stolen items, I pointed out that the disgruntled residents had the badgers to thank for the happy outcome.

I was committed to a year in my stop-gap home and made the most of the opportunity to explore what the environs had to offer. It was very much badger country, but there were relatively few foxes and deer. In my home valley I could expect to see deer and foxes every week of the year, but here in twelve months I saw deer only twice. In the evenings I would watch at some of the surrounding badger setts. One was in the corner of a wood and I had a favourite tree, a mossy oak with welcoming branches, for climbing and sitting in. Elevated perches have long been exploited by tiger hunt-ers and deerstalkers. Terrestrial beasts are not in the habit of looking up, and human scent is dispersed on the wind.

Another good sett was only 200 yards from my door. Derek had shown it to me once. Dog walkers passed by it every day but were oblivious of its existence. I straddled the barbed wire in the corner of a field and found myself at the top of a cutting in a lower field corner. I dropped down on to a ledge behind a tree, and dropped down again in order to check the sett by day. It was like stepping through a rent in the parochial fabric. When I watched there I perched on the ledge behind the tree where nobody could see me and tossed handfuls of peanuts and sultanas towards the entrances at the base of the cutting. Then the sheep would arrive to browse on the overhanging foliage. They stood in the mouths of the holes. I picked up small stones and threw them at their thick woolly flanks and, having got their attention, shooed them away with extravagant mimes. Eventually they would depart, reluctantly, haughtily.

The badgers emerged in late daylight. Because they were so close they seemed huge and a little scary, but they were only intent on eating their treats. Badgers never seem to give a moment's thought to the provenance of novel food. Foxes are more circumspect, approaching cautiously. The set-up for these vigils was so fortuitous that all I needed to do was slowly get to my feet while the badgers, two or three adults, were still busy and distracted, hoist myself on to the upper level and tiptoe away. They never noticed me leave – I could look down and see them still feeding. And I was home again in a flash. Rarely is badger-watching in the field this convenient.

Or I might walk north-west across the fields to what is the largest stretch of ancient woodland left in the borough. Here I discovered that a litter of eight fox cubs was occupying a sometime badger sett right next to a footpath. It is difficult to count eight active fox cubs but I watched them several nights running and checked and rechecked. The odd late dog walker, jogger, cyclist or horse-rider came by, and as the cubs heard them coming and went to ground so I also hid. Once the humans had passed, first I then the cubs came out of hiding, they to resume their play. Despite their *joie de vivre*, fox cubs are instinctively mute above ground while it is still light. When cubs have reached this lively stage, the vixen no longer dens with them, and the really special moments for the observer are when she turns up to pay them a visit. Then they crowd round her and beg for food at her mouth, and at some point she will suckle them since they are not yet fully weaned. While nursing her needy brood, the vixen seems to beam with pride.

Small fox cubs are tied to the security of a den. Badgers can be encouraged to hang around their sett by providing them with artificial food. But beyond these ties and centres of activity, the animals become much harder to observe rather than just glimpse. Of the two, foxes present many more opportunities for the watcher by dint of being higher at the shoulder and the fact that they are afield much earlier.

One fox family I regularly watched was lying up during the day in a belt of dense scrub between the railway line and a row of back gardens, as safe a zone as might be found in the modern world. There is a wood on the other side of the line. The five cubs, old enough to have left the den, would play on the tracks and I watched them through binoculars from a footbridge. They heard the trains coming and must have felt the vibrations long before the thundering monsters became a realistic threat. Dutifully, they all retreated into cover in good time and waited for the interruption to pass before coming out again. I doubt whether the train passengers could have seen them at all. In situations such as these, foxes learn wariness and the ways of the human world from an early age.

I could have been mistaken for a trainspotter, which meant that once again I was incognito. At such times as these I am caught in a blurred borderland between two versions of reality.

More than one fox may be attracted to a meadow that has been cut for haylage. The grass and wildflowers are left to dry out in windrows. Crows and magpies arrive first on the scene. They have learnt that the rows of cuttings contain mutilated prey, victims of the grim reaper, and anything still alive is deprived of its usual cover. There are easy pickings to be had and as dusk approaches the corvids are joined by foxes. Foraging foxes often have magpies in attendance. Opportunist following opportunist. The magpies may tease and provoke to distract a fox from its finds but they must be wary of ending up on the menu themselves. In uncut meadows, foxes spend a lot of time foraging for living invertebrates. Rodents and rabbits are comparatively hard work.

Unlike fox cubs, badger cubs are very difficult to observe when they are small, only appearing at the mouth of a hole long after dark. Knowing this, I obtained evidence of 'my' first litter remotely on trail camera. There were four cubs, which is a good complement. In the spring of the third year of my ownership there were two. This time I saw them in the flesh and got to know them quite well.

It was late May and very dry so I decided to support feed the badgers at their sett. Just before 8 o'clock one evening I scattered handfuls of peanuts, sultanas and dog and cat treats around the holes and at the edge of the field on my side of the sheep mesh which separates clough and pasture. Then I retreated to a respectful distance to watch. The badgers had been fed in this manner at certain other times so I wasn't particularly surprised when an adult soon appeared and got stuck in. While it was hoovering up the morsels, a small stripy head kept popping up at the main entrance until the whole animal emerged, sampled some of the nearest treats, then reversed into its burrow once more. Eventually it plucked up the courage to remain above ground in full view. Now a second small head and body went through a similar perform-ance. Each cub was already half the size of its mother (but probably not half her weight), fluffy and paler grey. Perhaps three months old. One, presumably the first to emerge and boldest, bullied the other, ragging it by the neck. It was little more than rough play but nevertheless an early indication of the social order to come. For half an hour well before sunset – indeed in dappled sunlight – I had good views of Mum and the two cubs.

I came every evening for four weeks during a period in which it remained cruelly dry, too dry for the family to reliably find worms to eat. Several times friends came to watch. The cubs weren't yet confident enough to emerge until after their mother was up, not that emerging in good light so early in the year is usual. For comparison, I drove to another location on my way home from watching at my sett and, using my car as a hide, rolled down the window. Ahead of me, the sunset sky was peaches and cream. A hare appeared on the other side of the narrow lane and, after some deliberation, ducked under the seven-bar gate, hopped around the car in that ungainly way they have when in a low gear, and went under a second gate. Here the pasture was close-cropped – by cattle and rabbits – and my view uninterrupted as far as the entrance to the quarry where the sett is. At 9.35, the light still passable, a rock I hadn't noticed before began to move and turned into a badger. This badger hadn't been enticed out early by the provision of an artificial food source, so was exhibiting a more realistic emergence time.

Back at my sett the bold cub, instead of tucking in straight away, would stand guard over the mouth of the hole and prevent the other from coming out. Until eventually giving in to its appetite. There would always come at last a peaceful phase when all three were busy feeding. One evening there came of a sudden the blaring of the village brass band on the march with the thumping of the big bass drum. The badgers took no notice. But when a woodpigeon clattered in the branches over their heads, they all dived for cover. No other adults joined the family while they were feeding; I suspect that the sow wouldn't have allowed them near.

Watching badgers snaffle artificial treats is all very well, but natural behaviours are much harder to observe. Witnessing the act of bedding collection, for example, requires a lot of luck. This task may, however, be performed in good light and the subject be so engrossed in its business that it fails to perceive that it is being watched.

Leaf litter, dead grass, greenery – and even, in some situations, human-sourced rubbish – is scratched and raked up into a bundle and the badger shuffles backwards to its burrow half clutching, half rolling its precious cargo. It might be imagined that this is a laborious process, but badgers in reverse gear are unerring and nimble. It is a bit like watching a skilful footballer taunting an opponent by showing the ball but dribbling it backwards out of reach.

Traces of bedding material at the mouth of a burrow or neglected bundles nearby are among the most conclusive signs of occupation by badgers. No other large British mammal indulges in this activity.

Relatively few setts lend themselves to direct surveillance. Some are obscured by vegetation while others are so exposed that the watcher would easily be detected either by the badgers or human passers-by, and many are on private property. The observer should familiarise themselves with the site by day and plan a discreet approach and departure. The locations of badger setts – and fox earths – are secrets that need protecting. Yet it is secluded setts, out of the public eye, that are most vulnerable to molestation.

As a rule fox cubs begin to appear as road casualties from early May and badger cubs towards the end of the month, although they are born earlier on average. The next live fox cub I came to handle had, like so many foxes in trouble, found its way into someone's back garden and was sprawled on the patch of lawn as if appealing for help. I had no backup on this occasion. Foxes can be very tricky to catch, so I thought I would enlist the help of the householder if need be. But in the event it was a simple business. The cub squirmed under a bush when I came close but further progress was impeded by a raised bed so I had it cornered. I bundled it in a towel and held its gaping jaws at arm's length before depositing it in the cage. At this point, being focused on the capture, I had made no attempt to assess the cub's injuries, merely taking in the fact that its coat was in good condition and its eyes bright. Buoyed up with the success of the capture, I didn't think to question why it had been so easy.

I set off to the vets in good spirits, thinking ahead as to what options there might be for the cub's rehabilitation. I'd keep it in one of my sheds if necessary, though at least one of the neighbouring farmers would go ballistic if they knew. The vet, however, wasn't so hopeful. I scruffed the cub while she put a tiny muzzle on it. Straight away she seemed concerned about the fox's inability to do this or that. She took it away to be X-rayed and it wasn't even necessary to sedate the cub at this point. Waiting for the vet's return, I could feel my upbeat mood draining away.

The vet identified the cub as male and pieced together a likely seeming scenario. A hindleg bore the hallmarks of an old fracture. This had been a serious handicap which probably explained why the cub was underweight and dehydrated. Only now it seemed there had been a second traumatic incident. He was exhibiting vertical nystagmus, an oscillatory effect of the eyes indicative of neurological damage – and his front legs weren't functioning. It was the end of the story for the cub, but we could sketchily reconstruct two earlier episodes involving vehicles.

There would be a series of rescued fox cubs and in each hopeless case I indulged in a short-lived daydream in which my radically altered routine revolved around the exciting challenge of meeting the needs of a demanding semi-tame, half-wild companion. I have read about a number of instances where foxes have become used to humans and their trappings but remain immune to our expectations and rules.

Yet, it should be borne in mind that infant mortality is high by design, so there is no real reason for sentimentality. Wild animals give birth to more young than are likely to survive. If some should perish due to hardship then the chances of the rest reaching maturity and continuing the line are improved.

Moreover, there is a danger that rescued young foxes may become too tame. They are then problematic, existing in a sort of limbo, unfit either for the wild or domestication. One of these lost souls passed briefly through the rescue centre on its way from being overfamiliarised with humans and dogs, in the absence of den mates, to be reconditioned at the fox sanctuary. When Pam entered its enclosure and bent down, the excited fox put its paws on her shoulder and licked her face. And when I encountered it on the other side of the mesh on a later occasion, it leapt up and clung on to the wire like a fruit bat – a flying fox – in a desperate bid for attention.

A number of old-school naturalists have raised fox cubs in order to study them. The pioneering fox researcher David Macdonald literally lived with foxes, hand-rearing a succession of cubs. When he 'exercised' the young fox Niff, he took her out on a lead, but only so that he could keep up. Over and over again, Macdonald emphasises that Niff was his tutor into the ways of the wild and he her pupil. Niff set the best possible example of what we would call 'field craft', the art of tracking and stalking. Writing around a century ago, H. Mortimer Batten suggested that foxes would make 'delightful pets' if it wasn't for their 'overpowering' smell. To my nose, they smell of fish, burnt rubber and drains.

A couple from the badger group, Maria and Jim, bravely took delivery of a consignment of six rescue foxes from Whitby, all of them sub-adults. The standard soft release procedure with young foxes is to keep them cooped up together for a couple of weeks, then to continue providing food for a time once they are at large. Maria and Jim had emptied their spacious garden shed of its usual paraphernalia, covered the floor in a thick layer of wood shavings and introduced dog baskets and bowls. But Maria told me that she couldn't bear to see the captives looking forlornly out of the windows from the work surfaces and so she let them out after only half the allotted time. They were now free to disperse into the adjacent wood, but two of the six showed little inclination to leave and skulked about the large garden. When I called at the house, I could see these two loiterers playing on the lawn from the kitchen window. The purpose of my visit was to receive instructions on the foxes' upkeep. Maria and Jim had previously booked a fortnight's holiday and now needed a fox-sitter to maintain the feeding regime.

They showed me round the premises, and when we ventured outside, the two scrawny playmates took little notice of us. Whenever one of the foxes seemed tired of the game and curled up on the leaf-strewn grass for a nap, the other soon pounced on it and their play resumed. The door to the shed was propped open. The smell inside was certainly overpowering. Maria picked up a dustpan and, demonstrating one of my duties, scooped up the latest leavings. The shed retained one of its functions as a latrine.

It was several days later when I returned to begin my stint. In the meantime I had been advised that all the foxes had now left the garden. Dusk was falling as I walked around the grounds distributing largesse in the form of raw chicken pieces, raw eggs, wet and dry dog food and strips of hide. I also topped up the bird feeders and filled bowls with peanuts for the local badgers. As I went about these tasks, I became aware that I was being watched. A fox was sitting patiently, like a well-trained dog, on the other side of the lawn. I threw it a piece of chicken. Now, a naturally wild animal would have interpreted this missile as offensive, but the half-tame fox knew better and received its prize with good grace.

This individual became increasingly familiar on my subsequent visits and I admit to spoiling him. During the day I would look out for items of roadkill, a squirrel or a woodpigeon, with which to cement our bond. When I arrived and the lone loiterer was waiting for me, I would toss the dead animal towards him before putting out the artificial food items. But so as not to miss out on the chicken, my foxy friend would trot off into the wood to hide the roadkill and come straight back. As the days passed he grew greedier and, following me around the garden like a shadow, would eat or hide as many pieces of chicken as he could before the warier foxes began to show up after dark. Of these others I mostly only saw the light from my headtorch reflected in their eyes. Which was as it should be.

6

Grass

As pandemic restrictions eased slightly, I moved back to the old valley. Some of its mills remain places of work while others are derelict and some have been converted into apartment blocks. Derelict mills provide homes for opportunistic wildlife such as jackdaws, converted ones for humans. It was to a flat in a converted mill that I now moved.

The apartment complex is made up of two main blocks and a row of terraced cottages surrounded by a communal car park. My four-storey block is divided into flats facing north or south. Since the mill sits in a dip the third-storey, north-facing flat I moved into is only on a level with the road that runs by, but there are mature trees on the far side to compensate for the almost ceaseless traffic. The south-facing flats are spared the road and have enviable views of trees and a field of pasture, but the residents must endure uncomfortable heat in high summer. The field contains rare-breed sheep and is regularly patrolled by foxes, but I have noticed that herbivores and carnivores take little notice of one another, even when the sheep have lambs. A brook runs in the bottom of the wooded ravine which separates the field from the mill car park. The ravine is both daytime refuge and safe corridor for foxes and deer. I chose to live here because I would no longer be in a village centre and would be surrounded by some of the wild animals which particularly interest me. Although they are much less apparent, neither are badgers very far away.

The first night I slept on the floor of the empty flat. The furniture was due to arrive the next day and my plan was to drive over to the rented place first thing to let the removal crew in. When morning came I looked out of the windows which face the road and saw that there must have been a minor accident. Three cars had pulled up and their drivers, standing about on the pavement, were on their phones. On the stairs I passed one of my new neighbours, who told me that a deer had been hit. I clattered down the last flight, hurried over and spoke to one of the drivers, who told me what had happened. A deer had run into the road and he had hit the one following, a youngster, which had got to its feet and scrambled up the bank on the other side, obviously in some distress. He pointed vaguely. It was up there somewhere, among the brambles. He and the other witnesses were trying to get through to a vet, the RSPCA, the police – but they were having no luck. I said I would deal with it and the drivers gratefully got back in their cars and carried on to work.

I picked my way cautiously through the dense undergrowth of the bank, not wishing to flush out or panic the deer, but when I spotted it from a few yards away it made no attempt to flee. To my mind, a roe deer fawn, or kid, is the most disarming thing alive. This one was about two months old, its cryptic spots no longer bright. All its features – eyes, ears, nose – were exaggerated, magnifying the kid's vulnerability. Its pretty head was like a flower on the stalk of its long neck.

It was my turn to get on the phone. As well as calling for backup, I rang the removal firm and managed to put them off till later in the day. This was my first experience of dealing with a live deer casualty and in the end I rounded up more support than I really needed. There were five of us closing in on the kid, which put up no resistance. It was surprisingly placid at first and let itself be caught and carried, but the situation changed once we had it in the car to take it to the vets. I was in the back of Philip's car with the kid, which I could now see was female. A hindleg was clearly broken. For the most part, injured wild animals are, mercifully, stoical and endure their pain and humiliation with what seems to us to be great dignity. At least in the case of adults. As soon as we set off the kid began, understandably, to give expression to its distress at being caught up in such bewildering circumstances. Now I needed Pam to help me hold the frantic creature down while it bleated piteously, thrashed and kicked. Pam had the back-end, me the front. We couldn't look at each other. I thought covering the kid's head with a blanket might calm it down but, in its struggles, it became entangled and couldn't breathe. It was also in danger of bashing its head. When it was exhausted, the kid hyperventilated, then began struggling once more. Philip drove in grim determined silence. The hellish journey seemed to go on interminably. Waiting at lights was almost unbearable. We looked out upon the everyday traffic on the everyday streets as if from the depths of a torture chamber.

At last we arrived at the surgery and almost immediately handed the kid over to a nurse. It was summarily euthanised. I felt terrible, stunned almost. I wished we had left it alone instead of tormenting the wretched creature. Its mother would be distressed, too, by the loss of its offspring.

Deer continue to cross or attempt to cross the road outside the mill. Another was killed in the year that followed and I have witnessed two successful crossings. In the absence of natural predation – lynx are specialist roe deer predators on the Continent – the death toll inflicted by cars has no significant impact on the British population of roe deer, whose numbers show an upward trend.

The first few months of a roe deer's life coincide with a time of plenty. As long as a mother holds a safe breeding territory there is little need for her to move her offspring to another site. The upshot is that very few spotted kids are numbered among road casualties. And by the end of summer the kids are swifter and wiser, better able to meet the challenges presented by a dangerous wider world.

I soon discovered vantage points within walking distance of the mill from which I could watch deer emerge into the open at dusk. The nearest is just around the corner where there is a disused nineteenth-century chapel whose car park, despite the sign – PRIVATE CAR PARK / CCTV IN OPERATION – is generally empty. The car park overlooks a rough meadow with a wooded ravine on its far side – the same ravine that runs behind the mill. Foxes and deer that have spent the day among the trees and understorey venture out into the meadow as daylight declines, foxes to forage or hunt, deer to graze. Roe deer are both grazers and browsers so a mosaic of woodland and grassland is ideal for them, though scrubland and large gardens are also suitable habitats. Apart from the empty chapel and empty car park, the back of just one house looks out over the meadow. An Alsatian lives there and if it is in the garden when I arrive it will bark at me standing there and my cover will be blown.

But if I am in luck and the coast is clear, I might witness something really worth seeing. A roe kid out of hiding, under the watchful eye of its mother, frolicking for the joy of being alive and free and safe. A buck in its breeding prime burning off excess energy by thrashing the tall weeds with its antlers. A fox perched on a wall, intent on the ground below, compressed like a squashed spring, then launching itself high in the air and coming down, forepaws extended, on to a vole in the grass; and, the rodent now stuffed in its jaws, trotting away. And, on a sharp winter's morning, the scruffy grass whitewashed with hoarfrost, a pair of bright orange foxes, dog and vixen, crossing the open space together. It isn't always necessary to travel to far-flung places in order to encounter wild animals behaving naturally.

The brook bypasses the mill and the chapel meadow, and carries on running down to the river. I walk under an arch of the viaduct and cross the canal at a lock bridge. On the far side of the river, where it meanders, is an even better sanctuary for large wild mammals: a long stretch of pasture which sometimes holds a few horses with a wood at the back, the whole area inaccessible to casual intents and purposes. I once identified six individual deer here: a breeding doe and two new kids and a dominant buck and two yearlings, male and female.

By leaving my newly acquired plot of land to grow wild, I hoped that it too would become a place of refuge for deer, but so far these animals have only appeared on site sporadically. The plot is a fifteen-minute drive west of my flat. I visit most days, every day when there are bird feeders to replenish or during a period when I am supplementing the badgers' diet. Sheep from the surrounding fields and moor come and go as they please. Often they knock the coping-stones off the tops of the dry-stone walls as a result of their incursions. I put the stones back as best I can but have no eye or feel for the art. But I bear the sheep no grudge; they come with the territory. When they get in trouble, I rescue them as I would any other animal. Sometimes sheep roll over on to their backs and then, like turtles, they are helpless, too fat and ungainly to right themselves. Eventually they will die in this position. They also get stuck in fences. But the most frequent intervention I make is to reunite lambs and ewes which have become separated.

Despite the sheep, the field, historically permanent pasture, soon became a riot of uncropped grass, thistles and nettles, the equivalent of a tramp's beard full of burrs. In summer it is alive with insects. The local tree society, a voluntary organisation run by genuine altruists, not only planted trees but were keen to dig a pond as well. The planted area which includes the pond is a discrete L-shaped strip higher up the hillside. The pond fills naturally from field drainage and moorland run-off and it has become a magnet for insects, frogs and birds.

The first spring nine clumps of frogspawn appeared in the pond as if by spontaneous generation. The next year there were too many to count. Then disaster struck. Frogs are in the habit of laying all their spawn together in one or two patches and when the tadpoles hatch their first meal is the jelly which had surrounded them as eggs and embryos. These hatchlings form a dense black mass at the water's surface, prevented from submerging by the clumps of vacated jelly beneath them. This seems to be a design flaw since they are vulnerable at this concentrated phase to mass predation and, in my case, a single pair of mallards ate the lot.

Despite my disappointment, I didn't hesitate to respond when called upon to rescue a brood of ducklings soon afterwards. Mrs Mallard had had the good sense to make her nest well off the ground – halfway up a cliff-face – but when the eggs hatched and the ducklings jumped down they were trapped within a closely fenced and locked backyard. It wasn't simply a matter of letting them out – not that a key ever came to light – since the duck would have attempted to lead her brood across the busy road to reach the river on the other side, there to be confronted by a further obstacle in the shape of a high wall – high by a duckling's standards, that is. I needed to catch the ducklings – and the duck.

I called Pam and asked her to pick up a net from the rescue centre on her way over – a fisherman's landing net with a long handle. I gave Pam the job of catching the duck while I would do my best to round up the little ones. The fence was low enough for us to climb over. There were eleven ducklings without any shade on this scorching hot day. In point of fact, I had been requested to rescue *a* duckling since only one had been visible at the time to my caller. It must have been the first of the brood to emerge into the light and take the jump. The brood split up into ones and twos and threes, some squeezing through gaps in the fence in their panic. I chased them round the yard and outside the fence, grabbing one or a handful at a time and stuffing them into a carrier bag. While focused on my all-consuming task, I was aware at a subliminal level of the rising and falling of the net as it slapped on the cobbles, the mother quacking frantically and trying to protect at least some of her brood. But when I had all eleven, it seemed that the duck had disappeared.

Not wishing to drag out the traumatic experience, I didn't pause for thought but crossed the road, climbed over the wall and scrambled down the steep bank to the river. Miraculously, Mrs Mallard was already there as if waiting for me to deliver her offspring as arranged. I tipped up the bag and counted the ducklings out. They took to the water with proverbial ease and the reunited family settled down in an instant as if nothing untoward had occurred, as though everything had gone according to plan.

For a long time I got a huge kick out of looking over my property: the field-turned-meadow with the wooded clough on one side and the plantation on the other, surrounded by moorland hills. This is *mine*, I said inside, almost disbelieving. I take pride in showing it off to visitors. Some of my old friends and acquaintances are out of shape and getting on, and aren't up to straddling the sheep mesh to get into the steep clough, which means they are denied the heart of the place: the hidden stretches of the stream and intimate perspectives of the badger sett.

'Rewilding' is pretty much a household word these days and, although I have never sought to apply that specific term in connection with my land myself, its currency seems to have lent a certain credibility and licence, kudos even, to the project among many of my neighbouring landowners and householders. To my knowledge, only one neighbour, a diehard whose family name has been associated with farming in the valley for five hundred years, has voiced discontentment. Once he told me to my face that he had never seen the field looking such a mess. I patiently explained that there were still plenty of 'tidy' fields all around and that mine was intended to cater for wildlife not livestock or pets. One man's mess may be another man's treasure. I know for a fact that he doesn't approve of me encouraging foxes and views my badgers with suspicion. That he culls foxes, I also know for a fact and feel certain that my badgers are only safe because I have the law on my side.

While the term is far too grand for my small-scale plans, I am happy to ride the wave on a turning tide. In truth, my plans amount to little more than seeing what will happen as more cover grows. If I don't interfere too much with what already exists, it is partly as I would wish it, partly due to lack of practical knowhow, and partly due to the difficulty of access. The land is steep and can't be approached by vehicles or machinery without permissive access from surrounding landowners – the pond was dug by hand. Besides, I don't drive a Land Rover. It is the ultimate aim of some conservation projects that ecosystems should be self-sustaining. And that's my excuse for letting the grass grow under my feet.

7
Remote

BEFORE I BOUGHT my first trail camera – or camera trap – I borrowed one from the RSPB and taught myself how to use it. Even a technophobe like me can get the hang of a new gadget if the motivation is there. A trail camera is an automatic device with an infra-red sensor which detects movement day or night; after-dark pictures are in monochrome, at night all animals are grey. In ideal circumstances the device should be strapped to a small tree on a so-called game trail, meaning a regular path used and maintained by wild animals.

The borrowed model didn't have a playback screen – and I didn't have a PC – so I used it on the stills setting. I would remove the SD (memory) card and insert it into my compact camera so that I could view and delete the pictures on that. I soon obtained my first good shots of foxes, badgers and deer, and these I had printed out in town. Foxes are wary of novel objects in their home range and are therefore typically camera-shy; badgers, being low to the ground, are not likely to notice a camera which is angled downwards even from waist height; but deer are very inquisitive and have a habit of shoving their faces right up to the lens.

It isn't likely that another person will come across your camera by chance, assuming it's been set up off the beaten track. On the other hand, if someone should happen across it, it is there for the taking. So I got into the habit of securing my camera to a tree trunk with a suitcase cable lock. Eventually, I had a range of models and would often have two or more in the field at once. Only once has one been stolen.

Derek took me to see an active fox earth at the back of a scrapyard. The scrapyard had always been a bit of a puzzle since its great mounds of rusted pieces of iron seemed to be fixtures rather than commodities yet the site boasted ample evidence of tight security. Far from constituting an eyesore, the collection of heaps had a certain purity of colour and form which transformed them into accidental works of art. Derek and I were in agreement that they were much more interesting than most of the exhibits on display at the Yorkshire Sculpture Park.

The den was outside the perimeter fence, but even so it was necessary to squeeze between slack strands of barbed wire and sneak through a dense windbreak without being seen. We were still trespassing. The entrance to the den was under a jumbled heap of broken concrete slabs like the aftermath of a localised earthquake. Cow parsley covered the ground but was trampled down around the mouth of the den. I took a camera from my rucksack, attached it to the bole of a tree 5 yards from the den mouth, removed a few twigs and tall weeds which might blow about and trigger the mechanism, and locked it in my usual fashion.

Two days later I returned, not to take my camera away just yet, but to check if I was getting good results in the position I'd chosen. It was a Sunday morning, quiet with warm spring sunshine. As I crept towards the den, trying not to make a sound, I was distracted from my purpose by the sight of a fox cub moving softly through the ground flora. Ten yards from the den, I froze in my tracks. Three more cubs appeared at the den mouth and stood watching the first cub. Deciding that the coast must be clear, the three joined the other and they all flopped down to doze in the open, in a patch of sunlight, using each other's warm bodies as cushions; another rust-coloured heap. But after a couple of minutes a restless cub got up and started to explore. It kept on coming towards me but, just when it seemed it must discover my feet and receive an unpleasant shock, it suddenly remembered its siblings, turned, and scampered back. Once they were resettled, I crept away in reverse, keeping my eyes on the cubs for as long as possible. I was pleased with myself for having had the luck to see them in the flesh and not disturb them. However, the next time I came to check the camera, I found that it had mysteriously disappeared. I could only hope that whoever had taken it was a good guy and had the fox family's interests at heart.

Sometime after I had moved into the flat I was contacted by the caretaker of the village primary school just up the hill. He thought that badgers had burrowed under the annexe classroom, and I agreed to meet him and take a look after the children had gone home. I didn't miss my old job and could take a smug pleasure in visiting a school as an outsider. There were several entrance holes in the flower-beds surrounding the Portakabin and, the caretaker explained, each morning it was necessary to sweep up the soil thrown on to the tarmac. I had brought two cameras along and, in order to monitor the situation, would need to set them up each day after school hours and remove them again before the children began to arrive in the morning. Since it was my local school, I didn't object to this commitment.

It turned out the tunnellers were a family of foxes. As far as I could tell from the video clips I obtained, there were four rangy cubs which really ought to have left their subterranean lodgings by now, but this situation was unusual. There was nowhere suitable in the school grounds for the cubs to lie up during the day, but there was presumably plenty of room beneath the classroom for them to move from one fouled recess to another, which would explain their ongoing digging activities. When the classroom was empty of children, the teacher could hear the animals moving about and scratching under the floor. And I couldn't help imagining what the foxes could hear over their heads: the stampeding of feet and the hollow scraping of chair legs.

I kept up my surveillance to see whether the cubs would move on by the end of the school year, but they never did. One afternoon I met the head who told me she was concerned that the foxes' tunnelling might cause structural damage, but she didn't seem to object to their presence per se, which was refreshing. At another school I visited, foxes in similar circumstances were perceived as a health hazard. As an end-of-term treat, the whole school, children and staff, watched and apparently enjoyed highlights of the extensive video footage I had obtained in the assembly hall. But I wasn't allowed to present the showing myself because there were still strict regulations in place regarding visitors to schools following lockdown.

A limitation of trail camera footage is that the action is confined to a narrow window. One way of keeping subjects in the frame is to provide food in a central location, but foxes will always take that food away – especially if they are aware of the camera. With foxes, I usually place the camera up high since they are not in the habit of looking up. Even when a deer carcass is the bait, foxes will drag it out of shot.

A second constraint is the necessary interval of delay between video clips. If the camera were on a continuous setting, the batteries and card memory would soon be used up by longueurs of stillness and emptiness. So the camera is triggered by a movement after which it runs for ten seconds, say, before shutting off to await the next trigger. The end result in this case where no bait was used was a collage of appearances and disappearances with cubs setting out purposefully beyond the wings of the stage only to return moments later and head back underground. I could imagine the children finding these abbreviated antics quite comical.

Camera traps are useful for surveillance in remote locations. Moorland badger setts are invariably among rocks but still vulnerable to forms of persecution other than digging. It is difficult for overstretched sett monitors to compete on rough terrain with gamekeepers riding quads. Perhaps our remotest sett is in a steep ravine sandwiched between grouse moor and upland pasture. When I visited it a few years ago it had been blatantly interfered with. One of three entrances was literally walled up with building blocks from an obsolete dry-stone wall. And, outside a second entrance, a snare was attached to a restraining post wedged between rocks. The snare was taken as evidence and I dismantled the blockage under police supervision. I was now anxious to obtain hard evidence of the continued presence of badgers in the sett.

In the absence of trees it was difficult to know where to attach a camera. However, I found a little cave nearby which I thought the badgers might use; inside, the camera would be out of sight and safe from the elements. The cave had two entrances, like a tunnel, and my only option was to prop the camera on the ground where it might easily be knocked over. If it was knocked over from behind then I would have no definitive proof of badgers. For bait, I had brought hard-boiled eggs from which I removed the shells so that the smell should be pervasive. I returned twice before getting lucky. A badger entered the cave at the further end, ate the eggs (now several days old), carried on coming and knocked the camera over on its way out.

Planting bait is an effective way of luring badgers and foxes within range of a camera. Deer may be drawn to the camera itself out of curiosity, but there is a better chance of obtaining good results if a territorial marker can be found. This is a sapling which a buck has frayed the bark from with its antlers and rubbed scent on. On more than one occasion when I focused a camera on such a target, I captured footage of several passing deer inspecting the marker. Deer also show curiosity towards their own dead, sniffing at a carcass in a way which suggests an attempt to comprehend.

I have been unable to ascertain whether foxes ever scavenge the corpses of their own kind. Crows certainly aren't fussy, and whenever I see a crow pecking at a fox carcass I think of an eternal circle: crow eats fox, is eaten by fox. That foxes sometimes eat crows is evident from the remains outside their breeding dens. Although speculation as to how foxes might catch them has given rise to some charming myths – in one Aesop-like scenario, a fox lies down and pretends to be dead; a crow sees a chance to scavenge and flies down to investigate, whereupon the crafty fox pounces – the mundane answer in our times is that crows are routinely shot by farmers and gamekeepers so their corpses are widely available. Doubtless, woodpigeons are acquired by the same means.

Although camera trap evidence can be revealing and is sometimes necessary, there remains a distance between viewer and subject. It is not the same as being there.

8

Food Chains

I EXPERIMENTED WITH MY badgers, used them as guinea pigs to test what artificial and natural foods they would eat. Once again I was gathering my own data and challenging received information. This made me feel like a field, rather than an armchair, naturalist. I accustomed the badgers to finding food in a level place at the bottom of the clough where I could spread out the various offerings. I placed the food items in plastic trays from ready meals and covered them with flat stones which only a badger could budge (note to self: have plenty of trays, must cut down on UPF).

A supply of dog food came my way from a dog rescue charity. The Clough Clan would readily accept most kinds of pet food: wet (tinned), dry (flavoured 'crunchies'), cereal or chewy treats; but they weren't keen on fish-based dry food. They ate eggs (raw or cooked), cooked but not raw potatoes, brown bread only if it was buttered or spread with jam, any kind of nuts or dried fruit. They left most raw fruit and vegetables. Neal emphasises the importance of windfall apples to badgers, but mine, which are familiar with this crop, show no interest. One of my regular callers who feeds badgers in her garden claims that they eagerly consume raw carrots, but it is the only instance of this preference that I know of. Everyone seems to agree on the animals' distaste for tomatoes. Another garden feeder spoils her nocturnal guests with boiled potatoes served in butter. Being less extravagant, I will give mine jacket potato skins which elicit only passing interest. Friends of a friend fed a regular visitor on cooked pasta and, because of its bulk and taste for Italian, christened it Pavarotti. Mortimer Batten, ever the eccentric, claimed that raisins soaked in sherry were irresistible to badgers.

But I was most interested in the badgers' reaction to carrion since this is a natural choice and something of a grey area. In my experience a fox, although it approaches with caution, will never pass up dead flesh. But what about badgers? I began with squirrel roadkill. Nothing doing. Then feathered roadkill. Same result. These items, too bulky for the plastic trays, were scavenged instead by crows and magpies. Elsewhere in the borough I had captured footage of foxes and badgers encountering deer carcasses. Foxes showed immediate interest in the bait but a passing badger would sniff at the corpse and turn tail as if somehow offended. I was frankly surprised when my badgers turned their noses up at an eggy-smelling woodpigeon squab which had fallen from a nest. Yet their conspecifics were eating raw chicken in some gardens. Different clans, it seems, have different tastes. Bearing all these findings in mind, I looked for evidence of what my badgers were eating naturally.

Where livestock dung lies on close-cropped pasture in the surrounding fields it is thoroughly raked over and the ground underneath torn up as, evidently, badgers have sought grubs, beetles and worms. 'Snuffle holes', made with the snout rather than claws, can occur almost anywhere the ground is soft enough. In summer, subterranean wasp and bumblebee nests are discovered and ransacked. These nests are vigorously defended by day but the insects are torpid at night, and badgers are well protected by their coarse pelts and thick skin. For late summer and autumn staples such as blackberries and acorns, they would need to travel further afield. Their well-used paths suggest that they mostly forage 'inland' – that is, away from the moor – probably visiting gardens as human habitations become denser. My rank field is a source of bedding material, not so much a foraging ground.

Some authorities frown on the practice of feeding wild creatures and I agree that we ought to think things through. Are we endangering them by luring them across busy roads? Are we presenting the food in a naturalistic fashion or encouraging 'lazy' habits? Are the food options healthy for them? On the other hand, it is difficult to care about wild animals if we can't take the opportunity to get to know them, to make some form of contact. We take so much away from them, it seems a shame not to give something back.

On an uncomfortably humid midsummer's evening – it was a Saturday and there was a very real potential for noise at the mill complex with so many windows likely to be open – I decided to sleep out on my land. As it turned dark around ten, I lay down on my mat in the compound where the ground is fairly even and I was safe from face-ripping carnivores. Almost straight away the width of the field between where I lay and the clough contracted and vanished altogether as the chuckling stream grew louder and snuggled up alongside. Otherwise, it was as peaceful as I'd hoped for, though my ears strained to identify the tiniest unfamiliar crepitation. Smells too were intensified in the dark – among them the rank odour of a dead sheep. The moon, less than half full, was bright by 10.30. My sleeping arrangements proved comfortable and the air was pleasantly cool here on the edge of the moor. I slept reasonably well.

I wasn't best placed to see anything much apart from the changing light, but I was treated to a vibrant dawn chorus.

This was introduced exuberantly by the burbling cry of a low-flying curlew at 3.15 when the light was grey and nebulous. Then came the main theme provided by song thrushes and blackbirds – of whistles and staccato chattering – growing ever louder in the trees of the clough. The breath of cool air on my upturned face came as a bonus to this free outdoor concert. I lay awake identifying minor contributions such as the twittering of the swallows which nest in the stable block further up the hill and the rapid ticking of a scolding robin. There was a faint background hum of flies like tense, subtle strings. From four the musical babble ebbed and thinned out and was gradually replaced by the plangent notes of woodpigeons. There came the insistent cawing of crows and the strident klaxon calls of cock pheasants closely followed by the deep thrumming of their wings. Blue tits and great tits joined in with the sounds of a busy typing pool. And lastly came the finches, the twittering and wheezing of greenfinches and the orgasmic spluttering of chaffinches. Later came the scrabble of a squirrel's claws on the roof of the nearest shed.

Whatever the badgers had got up to in the night, I had no auditory clue.

Much of the surrounding rough pasture belongs to absent landowners, which means I am free to roam beyond my boundaries. To the west is a derelict farmstead, a field's width from the clough. Plans to renovate the buildings ground to a halt with the pandemic and they continue to provide opportunities for wildlife in an otherwise monotonous landscape. A pair of little owls raised four chicks in the roof of the main house, but when hardcore was laid on the rough access track, thieves took advantage and stole the heavy stone tiles, and the disturbed owls have never returned. In the last couple of years barn owls have been recorded in the valley for the first time and one at least has taken to roosting at this site. A disused chimney is a good approximation of a hollow tree, a natural resource in short supply.

A third bird of prey which hunts over the rushy terrain is the kestrel. One evening during a long dry spell I stood on the edge of the rugged ground with the clough over my shoulder: I was watching two overworked kestrels struggling to hover in the still, warm air. Kestrels thrive on wind to facilitate their particular mode of hunting over open country. In the absence of a breeze they are forced to perch in trees or on wires, but neither of these options is available to them beyond the clough. Both birds were drooping in mid-air, exhausted-looking, almost vertical. Irritably, the larger female drove her mate to another quarter, but neither was having any luck. It was still light but a bat had appeared to my left and was hawking for insects up and down the edge of the trees. Then it strayed out into the open. The female raptor saw it immediately and drifted over towards where I stood. I had been standing still for some time now and she took no notice of me. She was intent on the flying mammal. She stooped at the flickering form and knocked it to the ground. And before the bat had time to recover, she swooped down and carried it away in her talons.

The staple prey species for kestrels and barn owls – the latter also hover but much closer to the ground – is the field vole, which is just as abundant on moorland as in meadows. These common rodents go through boom and bust cycles like lemmings (which are large voles). They are also vital for short- and long-eared owls and are the small mammal prey of choice for foxes, weasels and stoats. In years when the vole population crashes, owls in particular are unlikely to breed successfully. The food web often begins with soft rush, a hardy winter-green plant of boggy uplands. The voles strip the glossy green sheaths from the stalks and leave behind short sections of white pith in little piles like discarded matchsticks. The voles themselves, like small mammals in general, are rarely glimpsed. They are an unsung keystone species, lacking in glamour and presence yet carrying great diversity on their shoulders. The field vole has the distinction of being our most numerous wild mammal (around 60 million). But their overall biomass is somewhat less than that of the population of badgers, which, individually, can be five hundred times heavier. Badgers also predate small mammals such as voles and rabbits, digging up the helpless nestlings in their underground birthplaces.

To the east of my plot another rough field stretches and rolls towards another ravine. My badgers have a couple of outlying setts in the field and one in the ravine. Outliers are small satellite setts within a clan's home range that are not always occupied but are useful retreats for subordinates. The ravine is almost treeless: there are a couple of hawthorns and, at its secret head, a solitary oak. Because of the steepness of the ravine at this point, the crown of the oak and the crow's nest within it are on a level with the remote outlier, which is under a boulder. Beyond the far side of the ravine is a derelict shepherd's hut, a scree slope and a large rabbit warren. When I visit the old oak outlier, which is mostly active – I imagine a deposed crotchety old boar living out his last years in solitude – I am out of sight of human dwellings. Such places are to be treasured in our overcrowded age. The odd sheep dies here and rots undiscovered in the hidden stretch of the stream. One spring I saw a pair of newly arrived ring ouzels here. On summer evenings I might sit and watch the antics of the rabbits. Rabbits also occupy part of the clough sett, often emerging first during my watches. A number of commentators have observed that such cohabitees are for some reason immune to predation by the resident carnivores.

Something I learnt fairly early on is that badger setts don't exist in isolation; they are like nodes in a much bigger plan. There needs to be some degree of crossover between clans so that the gene pool doesn't stagnate. So where was the next main sett to mine? And the next one to that? One thing was clear: on this, the south side of the valley, mine was the furthest sett west before the moor. This particular moor is a sea of tussocky purple moor grass (*Molinia*) which for much of the year is straw-coloured rather than purple. It would be true to say that badgers couldn't subsist here on a sodden, acid substrate. Two options remain open to dispersing or philandering Clough Clan members: they can cross over to the north slope of the valley, running the gauntlet of the A-road in the valley bottom; or they can travel 'inland'.

The furthest sett west on the north slope was already well known to me. I had watched here many times. The main action took place in the bottom of a natural amphitheatre devoid of cover apart from clumps of rushes. I lay on the lip looking down among sparse nettles and thistles, sheep droppings and concrete rubble – tangible details which come into sharp focus when you're lying on the ground for up to two hours waiting for a natural history soap opera to begin. But it was sometimes worth it – like the time I brought a badger virgin along, an RSPB employee, and we saw four adults and three cubs in good light. At one point he hissed that he needed to take a leak but I intimated that if the noise didn't discourage the players, the smell certainly would. He would have to hold on. What we witnessed then felt like a rare insight since the cubs belonged to two mothers, the dominant sow with twins and a subordinate with a bigger only cub. The subordinate kept her cub away from the others, clearly apprehensive in the dominant sow's presence and fearful for her offspring's safety. The big cub, having no one else to play with, pestered its anxious parent for attention. The other two adults, presumably male, merely set out to forage well away from any potential domestic strife, the equivalent of feckless husbands off to the boozer.

The next main sett to mine on the south slope, beyond the outliers, downhill and across a B-road which marks the old turnpike route, is a Derek special. Only he would, in the first instance, have the bottle to explore a long-abandoned garden gone wild and fenced around with barbed wire – despite his stiff knees. Every badger group should have a Derek, obsessive and self-motivated. To protect badgers effectively, especially from the scourge of development, the full extent of their network of historical strongholds needs to be understood. Soon after I learnt of the existence of the sett in this secret garden, I obtained camera footage of four cubs residing there with their mother.

Derek's unparalleled working knowledge of the setts and associated landscape over a large swathe of West and South Yorkshire amounts to an unwritten magnum opus, an unusual and largely unsung life's work.

9

Dead or Alive

ONE FOX DIED on the back seat of my car. Pam, who was in the passenger seat, and I listened to its huffing and whining, the last sounds it made on this earth. It was a wonder that it had been still alive at all since it was split open underneath like an unzipped pyjamas case. Another made it as far as the surgery but exhibited agonal gasping – which meant the end had come – on the examination table. They seem such delicate creatures, not evolved to take the brute impact of a speeding machine. Most of the foxes I attended to were already dead, which was a mercy, often mangled, which was a shame. A fox passes between the wheels of a car and the rolling tarmac like an item of laundry through the rollers of a mangle, and the result is an animal-skin rug, bloody and full of broken bones; their eyeballs pop out in cartoon astonishment. There was one on the edge of the motorway, and therefore out of reach, which I passed several times, whenever I returned from a weekend shopping expedition in Leeds. Each time it was a little thinner, a little more weathered, until it resembled a moth-eaten fur stole which might blow away in the draught of the slipstream.

After Covid, I stopped going to Leeds. I had become a little agoraphobic. The thought of bustling city streets, and of the merciless motorway, brought on panic attacks. I was too busy anyway; the death toll on the roads intensified as the volume of aimless and would-be busy and pleasure-seeking traffic increased still further. People had to get out of their homes and drive, even if there was nowhere particular to go. One weekend in November I picked up a dog fox on the Saturday morning and, on the Sunday, a vixen, probably its mate, on the opposite side of the same stretch of road. Call me sentimental, but I laid their broken bodies together beneath a bush.

Not only did I stop going to Leeds, I stopped needing to go. I was undergoing a sort of existential crisis. There didn't seem to be any books I wanted to read any more, or films I felt like watching, or music I could bear to listen to. It might just be my time of life and dissociation from current trends, but all I wanted was to be outdoors, and alone, and even if most of the animals – the larger mammals – I encountered were dead, I still wanted to be with their kind.

Like the fox that passed away in my car, it is surprising sometimes that a casualty should still be breathing at all. Another example was a badger which had been discovered next to the railway track during a routine line inspection. I have no way of knowing how many animals are killed on local railway lines. Only very occasionally is a casualty, visible from a bridge, reported, but venturing on to the line is not an option since it incurs a heavy fine.

On this occasion, we – myself, Pam and Philip – were escorted on to the line between trains by railway officials dressed all in orange. It was a single bidirectional track and the badger was 50 yards from a branch line station. We were instructed to walk on the ballast of limestone chippings, not on the wooden sleepers. The casualty was a big boar and he was tangled among the brambles at the side of the track. The boar was clearly partially paralysed since his back end was twisted belly up while his front end was face down in a bid to run away – like two halves of the same body with opposing wills. No one was in any doubt that this badger had been hit by a train and was, miraculously, still alive. Not just alive, but feisty at one end. The crazed creature tried to bite any foreign object which came within its reach. While Philip slid the open end of the cage over the badger's head and shoulders, Pam and I levered its inert back end forwards.

The X-ray showed a clean fracture high up on the right femur, the two sections of bone shockingly displaced. The badger had no 'anal tone' and there was blood in its urine. The affected leg, the vet declared, had 'felt like jelly'. All these details were academic, however. We had known all along what the badger's fate was to be. To say, idiomatically, that a body, living or dead, appears to have been hit by a train is to use one of our ultimate expressions of violence.

Despite, or because of, being lightly built, of the three common large wild mammals in my area, foxes are the only ones which are capable of surviving a broken leg and getting by afterwards. Badgers are too short-legged and heavy, roe deer too long-legged and heavy, to manage on three legs, but foxes, like similarly disadvantaged pet dogs and cats, can still function, partly because of the ratio of weight to height of their bodies but also due to the availability of food – including human handouts (inadvertent or intentional) – to a habitual scavenger. Limping and functionally three-legged foxes are well-documented inhabitants of human environments.

The boot of my car – one of a succession of small hatchbacks – has become a cabinet of curiosities with everchanging exhibits. As far as possible, I dispose of corpses close to the scene of the accident. In this way they are dispersed throughout the region. It also means that living animal companions can find them and perhaps understand that they are dead. A bereaved vixen, for example, must find a new mate. But if there is nowhere suitable nearby, if the location is exposed or built-up, then the body must be removed to somewhere more appropriate: one of my regular discreet sites, where the deceased is given an ostensible 'fly-burial'. On these occasions I am aware how easy it is to get away with fly-tipping, but these biodegradable packages are an important resource for scavengers and decomposers in a contradictory world that is polluted and poisoned on the one hand while overly tidy and hygienic on the other. Badgers like to eat beetles, among other things. Some beetles like to eat dead badgers.

You might not think a full-grown deer would fit in the back of a small car, but roe deer are not that big. I am used to taking calls about 'baby deer' casualties only to find they are mature individuals. Baby deer do not have antlers for a start. An adult roe is about the size of a large dog and weighs around 25 kilograms. Before I came to concentrate on deer I usually relied on assistance from someone with a bigger car for their removal. Once, Pam and Philip turned out on their way back from the weekly big shop and had to transfer the bulging carrier bags to the back seats and footwells before we could get the deer in. I admit my car sometimes smells bad, but I see it as a utility vehicle, not for show.

Where possible, I enlist the help of an inquisitive passer-by in lifting a deer into the back, but when no assistance is forth-coming I have found that I can manage. Many of these impromptu helpers are surprised to see a deer lying dead on the pavement in a built-up area, but from my perspective this is not unusual. Under cover of darkness, deer often enter perilous zones to get from one place of refuge to another, between isolated fragments of woodland, say. In mid-November a mature buck had been hit upon exiting a gap in a cemetery wall. His antlers had already been cast. He was intact – no legs were broken – but a pool of bright thick blood, like the erstwhile contents of an overturned tin of paint, was congealing on the pavement. It had, it appeared, poured out of one ear. There was no one around to help so I wrapped the buck in a doubled-over sheet and dragged him to the back of my car before taking a breather. I had dropped the back seats in readiness and spread an old blanket on top. Now I grasped both pairs of legs, lifted the deer clear of the ground and swung the body in. I had only to open each back door in turn and pull him forward a little. To finish up, I took fistfuls of wet fallen leaves from the gutter and mopped up the worst of the blood.

Another fine buck – it was midsummer – antlered this time and in light chestnut breeding pelage rather than winter's grey plush, had been struck with such force that concerned locals saw fit to display his body, slumped and broken, in a wheel-barrow, where it suggested a guy with the stuffing knocked out of it. If it hadn't been for the makeshift bier the deer's viscera would have spilled out all over the place. Derek was on hand to help me lift the laden barrow over a wall and I trundled it down the narrow path on the other side, out of range of dog walkers, and tipped the grisly contents out behind a bush. One hindleg was shattered and twisted and a foreleg was broken too. Compact coils of gut lay in a heap like a display of blood-sausages. The empty barrow was streaked and splashed with red. The buck's proud head and antlers had escaped damage so that the carcass was a chimeric combination of beauty and horror.

Outcomes such as these are the result of accidents, but let's hope they also act as wake-up calls. A deer can step out of nowhere, a flash of wildness unleashed on our everyday, taken-for-granted version of reality. Its abrupt, in-your-face appearance comes as the first shock, its inner vulnerability as a second. Different layers of reality are revealed. No one wants to hit a deer, and we don't like to be reminded that our surface selves, like theirs, conceal involuntarily working parts which, like those of a clock, will one day be broken and cease to tick.

Persevering with my unwholesome pastime of revisiting the dead in the afterlife, I chart the dismemberment of deer carcasses by foxes; or sometimes they are picked apart by buzzards and corvids, or even wholly consumed by maggots. A buzzard is best equipped to break into an intact corpse with a hooked beak like a can-opener, but foxes, unless the body is already broken, will enter, to put it politely, at the rear-end. Foxes will scavenge rancid badger carcasses when the time comes, finding fresh specimens tough going, but they are so fond of venison that they will make short work of most deer. A family of four foxes – the two parents and two sub-adult offspring – consumed a mature doe in five days, leaving nothing but bones and the spilled khaki-coloured contents – part-digested green stuff – of the rumen, the first stomach. The legs are yanked off in a series of tugs, the ribs are bitten off short, even the face and skull are gnawed. A friend who monitored the fate of a deer carcass on trail camera found that it was demolished by a loose pack of farm dogs.

Should I choose to claim the antlers, I cut through the exposed pedicles – the permanent bony stalks beneath the skin – with a junior hacksaw. A farm and equine vet demonstrated that they can be whipped off much more quickly with cheesewire, but it takes a second person to steady the head. Roe antlers, at about 20 centimetres long, are modest but nevertheless beautiful, comprising a brow point, rear or middle point and top point. They are further distinguished by dense 'pearling', especially towards the base of the 'beam' and round the bulbous 'coronet'. This covering of bony tubercles provides a rough texture like that of coral. Further texture is added by more or less vertical 'gutters' like fluting. The over-all effect is of a three-branched candelabrum holding guttered candles, or a three-fingered hand of glory. The collector should refrain from bleaching the antlers since they would lose their resinous, stained finish; they should merely be picked clean of dirt with a sharp point and wiped with a cloth and warm soapy water. They are tactile objects with the feel of tropical hardwood or Bakelite.

After the old antlers are cast in autumn, new ones develop during winter. The growing bone is clothed in so-called velvet and supplied with blood and nerves. At this stage the antlers are living, sensitive extensions; the nerves give the deer kinaesthetic awareness of its extremities so it may avoid obstacles. After the velvet is shed and the blood and nervous supply atrophy, the mental map remains like phantom limb memory. The bone is now dead, hard and ready for action. Modest they may be, but ornate antlers can make unwieldy weapons while a roe's fully developed headgear is lethal on the end of an energetic assassin. The behaviourist Konrad Lorenz made no bones about the inadvisability of keeping a roe buck captive, and roe specialist Frank Holmes recorded a vicious attack on a postman riding a bike, presumably mistaken for a rival. That said, I have yet to feel in any danger in close encounters with either injured or healthy deer. It should be noted that the purpose of antlers is not primarily to inflict injury, otherwise they wouldn't be branched. They are designed to interlock with those of a rival so that, should they engage in combat, the contest becomes a trial of strength. However, if an opponent is forced to the ground, it may be gored as an extra measure. If a deer possesses anomalous unbranched antlers, it is known in deerstalker jargon as a 'murderer'.

Yet deer casualties remain a tricky proposition, not just because of their size but because of their (short-term) survival rate. About one in three are not killed outright in collisions. And then there is the problematic, traumatic aftermath of the accident to be dealt with. Happy endings are exceedingly rare. Broken legs, punctured lungs from snapped ribs and spinal injuries are the norm and the wretched beast must be dispatched by a vet or a marksman. In one instance an experienced pest control officer, who happened to be passing, had suffocated a doomed victim before I arrived on the scene.

Hardly a day goes by when I don't find occasion to curse another road user – or two – for reckless or impatient driving, but I have seen evidence of collisions which were perhaps unavoidable. A mangled buck, the hide flayed by friction from one haunch, lay 50 yards back from where a van, its front end also mangled, had pulled over and was awaiting recovery. The driver had dash-cam footage of the accident. It was a fast road with a deep scrubby verge at the bottom of a planted embankment. Two deer shot out of cover into the road, in front of the van, without warning. The leader was a doe and although she just made it, if there had been anything coming in the other lane as well there would have been mayhem. On the other side of the road a crash barrier separated the pavement from a second embankment above the parallel motorway. There she must have fled. But the closely following buck, nose to tail, knew he wasn't going to make it as soon as he burst on to the tarmac and his legs buckled under him *before* the impact. The van went over him. Another unfortunate driver said that the experience was 'like hitting a great big pothole'.

Still-breathing casualties vary from the semi-conscious to the semi-mobile. One of the latter evaded capture despite both forelegs being shattered at the 'knee' (actually the *wrist*) – until he was cornered on a building site and hastily fenced in with movable sections of Heras block-and-mesh panel, then shot in the chest with a double-barrelled shotgun, first from a few yards away then point-blank (a firearms officer will dispatch an immobile casualty with a single shot from a pistol to the head). Most immobile casualties aren't difficult to restrain and stroking them – which comes instinctively – seems to help calm them down (this is definitely *not* recommended in the case of an injured badger or fox!). If a deer casualty has to be transported to the vets when no vet is available to come out, it becomes necessary to bind the legs. If euthanasia is by lethal injection, the tainted corpse must be incinerated. A large dose of urine-coloured fluid is injected into the deer's jugular. The end comes swiftly.

Sometimes the victim gets away. Carrying a broken leg, it will die from infection or the inability to function normally and meet its needs. I have trailed individuals which have escaped and come upon others with old fly-blown wounds and on their last legs. Sometimes I trailed splashes of blood, sometimes slots (hoofprints), sometimes the visible hobbling animal. One of these limped ahead of me through the shallows of a river which I concluded was a safe corridor in regular use, connecting two green spaces, but a busy road had to be crossed to access the river. I won't pursue an injured deer on its own ground; by that point it has suffered enough distress for one day. I elect instead to respect its chosen course of action and leave it to die in peace where it best belongs.

10

Cheese

I GRAVITATED TOWARDS DEER for a number of reasons. There wasn't a deer group, no one had a duty towards these animals. I could make the responsibility my own. Deer – specifically roe deer – seem to me to be the most mysterious and wildest of our larger beasts. The least known and least considered. I wanted to get as close to them as I could in order to still feel alive. And I needed to be defined by something *different* and exciting. To say I was a retired teacher, to myself or anyone else, was never going to be enough. I was a self-styled 'conservation and animal rescue volunteer', but that was still a bit vague.

As with badgers when I first set out, there were questions to be answered. How many roe deer are there in the borough? How many are killed on the roads? Are they distributed evenly or locally? How are they perceived by people from different walks of life?

On the printed page, roe deer are inseparable from the words 'graceful' and 'elegant'. In its own time, standing alongside a mesh livestock fence topped with two taut strands of barbed wire, a roe deer will clear the barrier from a vertical take-off with contemptuous ease, like a cat, as if in slow motion. But they are always on the lookout for gaps in our defences. School grounds, invitingly open and green, and invitingly empty during the holidays, are surrounded by high, impenetrable fences. But where the ground is uneven there are inevitably gaps underneath. It shouldn't surprise us that foxes and badgers can get in, but roe deer squirm underneath as well. The species has found ways of infiltrating our domain and live out an almost extra-dimensional existence in our midst.

Not so many years ago, like most other people, I didn't see them, was hardly even aware of their presence. It is tempting to conclude that they have become more common in the interim, but there are other factors at play. The exponential rise in the sharing of information giving rise to greater awareness; the continued, accelerated expansion of human encroachment on wild spaces; concomitant increases in car and dog ownership; and the progress of my own obsession. Any or all of these causes could make the deer more *visible*, but their perceived abundance may be an illusion.

My first visual impressions of roe deer came from illustrations in books. None were finer than those of Archibald Thorburn. My parents had bought a copy of *Thorburn's Mammals*, a 1974 edition of his 1920–1 work *British Mammals* with an introduction by David Attenborough (a household name even then). It was a coffee-table or rainy-day book, no one in our house actually read it. The colour plate of the adult pair of roe deer against a green wash was lovely, but there was something even better: the black-and-white tailpiece of a dead buck accompanied by a perched raven, chief mourner, undertaker, demolition worker. The buck's eyes were still intact, his antlers sharply recurved – the drawing seems to be based on the same model as that of the colour plate. I was fascinated by pictures of dead things from an early age.

The first time I handled a dead deer was decades later, but before I joined the badger group, before dealing with large animal casualties became routine. I was house-sitting for 'Our Kid' in North Yorkshire, tending to the hens, carrying out other minor duties, but mostly free to roam in a largely flat landscape of vast arable fields, comparatively deep woods and endless thick and high hedgerows. It was fabulous badger, otter and roe deer country. There were barn owls roosting in the isolated barns and bright yellow wagtails among the green potato foliage. I could walk out of the picture-postcard village and carry on going all day, between farmsteads, hamlets and woods, hardly encountering another soul. It wasn't all idyllic, mind. There were the pig farms with massive barns full of hundreds of cooped-up, stressed-out pigs and the fractious squealing was maddening and I wondered how the farmers and their families could stand it all day every day.

I have come to understand that roe deer are common in West Yorkshire; in North Yorkshire they are very common. But don't *expect* to see one every time you step out of the door. It is in their nature to be elusive. Besides, 'common' is a relative term in reference to large wild mammals.

On the first day of my stay during a summer heatwave, I set out north from the village, a settlement which ends abruptly, the speed limit on the through road jumping from 30 to 60 as one enters archetypal farmland-countryside with deep verges but no pavements. There. on the first bend ahead, a large animal was lying in the gutter, a brown dog perhaps. But as I drew close, buzzing with the excitement of discovery, I saw that it was a beautiful chestnut doe, intact, bloodless. There was something unnatural about the way her legs were splayed every which way, as if she had gone down on ice, but they were unbroken. I took these slender legs and dragged her up on to the verge into the high herbage. Over the next ten days I visited her daily, as if paying my respects at an open grave, by the end of which time maggots alone had stripped one haunch down to clean bone. And every day I saw her living counterparts, especially among the cereal and legume crops at dusk, helping themselves.

The deer have *space* up there. Here where I live they are often compromised by the fragmentary nature of the environment. There is always the worry of flushing them out of scraps of cover and towards danger. A panicked deer is not a good decision-maker.

I have also been a regular house-sitter – and dog-sitter – over ten years now, for an ex-teaching colleague and her husband in another part of my borough. The dog is a Border collie, one of the few breeds I have any time for. Towards the end of the long dry summer of 2022, the dog now decrepit, heavy and a bit senile, as well as deaf to the world, I took him on a short walk through the nearest wood, about all he could manage. The hardest part was getting him to his feet at the outset since his arthritic legs were so unsteady. But once off the lead he tottered away like a clockwork toy, ever in danger of keeling over, taking his own inexplicable course and impossible to call back. In the old days he would tree a squirrel and sit at the foot of the trunk barking.

Now he had entered the bed of a dry stream and was following it, meandering downhill. I watched him with some amusement before attempting to catch up and steer him back on to the path. By the time I joined him, I saw that he was sniffing with great interest around a fresh deer carcass, a handsome buck, which was buzzing with flies, flashing in the sunlight like a horde of miniaturised paparazzi crawling over the corpse of a celebrity. I praised the old dog for his unanticipated cleverness – there was nothing wrong with his nose – put him back on the lead and tied him to a tree while I examined the deer. Turning it over, I saw that no legs were broken; there were no scuffs or any de-gloving normally associated with road accidents, but its belly had been torn open. The violation of his body wasn't the result of scavenging post mortem, but a clue to the cause of death. Previous experience told me he had been run down by dogs. It was an example of another way in which deer meet their end in the Anthropocene. Others are shot, drown in the canal, become inextricably entangled in fences, or are hit by trains. But road traffic is by far the biggest killer.

Dead animals don't move. You can get up close and go on looking, and touch, and marvel. And notice things it would be difficult or impossible to see in life. The violet gland of a fox (indicated by the dark patch near the root of the tail), the metatarsal swellings on a deer's hindlegs, the long claws on a badger's forepaws and the subcaudal pouch under its tail. Now you can sex a badger or a fox positively and estimate the age of a deer from an inspection of the lower incisors. Only now will you discover that a deer smells of ripe cheese.

A more socially acceptable pastime of my retirement years is to scour charity shops and antique stalls for bargains. Even if I don't buy anything, I enjoy the pursuit, and the unpredictable clash of jumble on offer is always amusing. One thing I had set my sights on was a good specimen of a red deer antler at a reasonable price. It needn't be a pair, but it must be a six-pointer (from a 'royal' stag): brow, bez, trez and a 'crown' of three points. And it must be whole, that is naturally cast, not sawn off. Well, I finally realised that goal – after several years of keeping an eye open – at the covered market in town. In the meantime I had come across some shoddy excuses at extortionate prices, but here was a good one among a miscellany of junk, a job lot cast up on a beach after a storm. And it was unbelievably cheap. The second point (bez) was damaged, perhaps in a real live fight, but otherwise it was very fine. It now rests in a corner of my sitting-room crying out to be handled and admired, a true object of desire.

On an evening walk in my home valley in mid-July, taking in a succession of hill farms and the sixteenth-century manor house, I was astonished to spy a red stag on the far hillside. It was around 9 o'clock, the light still good, and the stag was browsing on bilberry foliage having presumably emerged from the remote plantation on the crest of the hill. Local place names include the appellations 'Stag', 'Hart' and 'Hind', so there must be a history of these animals in the area, but this one was a living relic. It was a magnificent specimen to boot, a royal, and on that first occasion, although he was more than 100 yards away, he looked up and saw me as I crossed a field, but held his ground. I needed a witness, so another evening I took Pam along and we sat in the long grass and waited, and eventually he appeared, stepping through the dilapidated dry-stone wall, a collective vision after all. I saw him again the next summer, but I suspected he must travel a long way over the moors (south into Derbyshire, perhaps) to participate in the autumn rut.

Roe deer are unusual – very unusual – since they are the only hoofed mammals *in the world* to practise delayed implantation, a breeding strategy otherwise only known in some mustelids (in Britain, in badgers, pine martens and stoats), bears and seals. This means that the roe rut is in high summer although the young are born in the usual late spring season for deer. This anomalous behaviour also means that the buck's antler cycle is out of synch with other male deer.

There are more ways in which roe deer differ from other deer. The pale rump patch of a deer is called a 'target', the tail is a 'single'. But a roe deer does not possess an external visible tail; the doe, however, has a 'tush', an inert tassel of long hairs. Before a roe deer beds down in its woodland 'lair', perhaps to chew the cud, it scrapes away the leaf litter with its hooves like a dog fussing over a blanket, leaving a diagnostic bare patch behind when it moves on. Although a non-herding species, roe form small aggregations in winter. Thus far, six is as many as I had seen together in my region. This particular group, observed over several days, consisted of a single buck and five does, but the latter did not represent a harem. A winter group may contain more than one buck. Sex hormones are at a low ebb and there is little rivalry at this time of year. Then, in the new year, I saw a group of eight still together in early April: a mature buck with clean antlers, two mature does, two young bucks in velvet and three young does. During the rut, when a buck pursues a doe in order to mate, she may lead him in circles around the same tree or bush so that a ring is worn in the ground, or a figure of eight if two trees or bushes are involved. Roe deer emit gruff barks of alarm when disturbed, loud and quite intimidating.

Since roe are the only type of deer likely to be encountered in my region, it is only necessary to distinguish the signs they leave behind from those of sheep. Deer and sheep are cloven-hoofed mammals. The two cleaves of each foot leave slots in soft ground. The hoofprints of the two animals are quite easy to tell apart. Those of roe deer are narrower and more pointed, overall more elegant in outline. Sheep prints are blunter and more rounded. Deer are more likely to vary their pace. When they run or jump, the cleaves splay and the two dew-claws (vestigial digits) come into contact with the ground, leaving impressions in mud, snow, sand, etc. The dew-claws of a sheep are placed too high up on the leg to ever leave any marks.

The droppings of the two animals are much harder to distinguish, especially if they are scattered, but sheep droppings often clump together (resembling soft hand grenades) whereas roe deer droppings typically don't. Individual deer droppings are called 'fewmets', heaps are 'crotties'. Male deer, including roe bucks, fray the bark from young trees with their antlers to mark their territory. The trees are slightly damaged but not killed. Where bramble leaves have been neatly snipped from their stems along woodland trails, you can be sure that deer have passed by. Bramble and ivy are two roe staples; being wintergreen, they are available all year round. Roe deer are both browsers and grazers.

There is a rich, albeit archaic, vocabulary specific to deer. These words, a number of which I have already drawn attention to, are known as 'terms of venery' and are mostly known to hunters and stalkers. The Scot Frances Leviston gave her evocative poem about a road-killed deer the enigmatic title 'Humbles' but all becomes clear when it is known that 'humbles', or 'umbles', refers to a deer's entrails. Once when a deer carcass was reported to the local police as having been 'gralloched' (gutted), the officer on duty recorded *garrotted* so that I was expected to believe a deer had been strangled by a maniac.

II

Action

A FTER A YEAR in which I had dealt with over two hundred large mammal casualties and five years of concentrating on badgers, I elected to take a break. I changed my phone number and left the badger group. However, a couple of deer casualties still came my way and by spring I decided it was time to reinvent myself as the Deer Man. By the end of the year I could claim that, for no apparently logical reason, DVCs (deer–vehicle collisions) seem to occur in clusters like the arrival of proverbial buses. The most dramatic cluster involved eight deer in six days at the end of April and the beginning of May. After a cluster, there might come a quiet spell of up to several weeks. The longest gap was five weeks between late August and the end of September.

As well as dead or injured – usually mortally wounded – deer, I respond to reports of 'trapped' individuals, deer which have got themselves into awkward situations from which they are seemingly unable to extricate themselves. An example: a female kid, perhaps six months old, separated from its mother and sibling, stranded on a narrow strip of private ground between a fence on one side and a wall on the other. This was in a built-up area and it was a busy time of the day. Sometimes I'm not sure what it is my callers expect of me, whether they assume that I carry a gun which fires tranquilliser darts, for instance. The elderly woman in this case, having stopped yet another passer-by and explained the deer's predicament, boasted, 'This man's going to rescue it.' However, I insisted on a calm, quiet and discreet atmosphere while I carefully explored the options open to us. Panicking the deer would be unavoidable, but the risk of doing so must be kept to a minimum. I asked the woman to stand guard while I surveyed the surrounding area and tried to figure out where the kid's family had gone and a safe route for her to be reunited with them.

I had little doubt the kid could clear the wall if it put its mind to it, but it ran up and down the high fence instead, poking its nose into the mesh, vainly looking for a gap. Over the wall was a care home. I knocked on a side door, the door to the kitchen. That was when I learnt about the two other deer, confirming my suspicions as to what had happened. One of the cooks showed me a brief clip of the deer on her phone. They had been in the yard but, alarmed by the awakening urban environment, mother and brother were long gone. The direction they must have taken was clear enough: the quiet back street rather than the busy main road out front. They were, I decided, headed for the park.

I considered cutting a hole in the fence, but although the former churchyard – now a spacious private garden – next door offered the sanctuary of plenty of evergreen cover, the owners kept dogs and the kid would be no closer to finding its kin. So, the elderly woman making more noise than I felt was necessary, we spooked the deer into jumping the wall and I tailed it through the grounds of the care home and along the back street. Fortunately, there was no one about, just a couple of startled cats. As it progressed, the young deer tested gateways and gaps on either hand but each time returned to its course along the street. I was concerned about what might happen when it reached the end where there was a perpendicular busy road – the public park was off to the left along the pavement – but instead the deer veered off into an industrial park where it continued to explore the perimeter fence. I called in at the office, explained the situation and left my number. On my way out I could see that once the gates were shut at the end of the day the deer would be trapped once more, but there was nothing further I could do. Deer are not accounted for in such an environment. Almost no one expects to see them on the street, and the deer find themselves in a maze of blind-alleys. I had to trust in this individual's ability to sort itself out when left alone, but it would need a certain amount of luck to negotiate further possible barriers and obstacles to its goal. All I can say is that I heard nothing more, and I told myself that no news was good news.

Another kid, male this time, had been caught in a fence. Deer entangled in fences often die, even after being freed, from a combination of exhaustion and psychological trauma. This one was doomed due to a further complication. Something had begun to eat it alive. The kid's groin area was gaping and gory. A farmer had found it and laid it down on straw in a vacant outhouse. His wife, kneeling down, was trying to comfort the suffering animal. It was October and the window and rafters, being festooned with swags of ancient dusty cobwebs, seemed decked out for a Halloween party. There was no doubt in the farmer's mind that the assailant had been a badger. I had heard a number of tall tales about badgers eating their victims, including sheep, dead or alive, inside out from the back-end. I had even been told that they killed cats. I didn't know what to say, apart from I hadn't seen anything like it before. I should have liked to have examined the scene of the discovery for evidence, but my host was really only concerned for the stricken deer. As soon as I laid eyes on it, I told him it must be euthanised without delay, and rang for a vet.

A yearling doe was still in summer pelage in the middle of October. All the other adults I saw that month had turned. She was semi-mobile and had been spotted by a couple out walking their dog, but now she was cowering behind a screen of brambles with a mesh-covered five-bar gate at her back. She was on the verge of a quiet narrow lane, so quiet that not a single vehicle came by all the time I was there.

When the vet, a young woman I had met on a previous errand of mercy, arrived, we closed in on the doe in a pincer movement. The doe tried to get up and let out a hoarse bark of alarm. The bark of a roe is quite startling, but in this context it was upsetting to hear. It was the voice of despair, of prey captured by predator. Once we had laid her on her side, she ceased struggling – I had my arms wrapped around her shoulders and forelimbs – and the vet was able to examine each leg in turn. We shared our observations on how shockingly thin she was. A hip was swollen and sore and the vet thought the joint might be dislocated. We gently turned her over to check the other side, then lifted her carefully to her feet. The doe was unsteady on her back legs, like a newborn. After some thought, the vet revised her diagnosis to a days-old fractured femur. It was hard to be sure without the benefit of X-rays, but the injury was evidently incapacitating.

'I think we're going to have to let her go,' she said.

This could be interpreted in more than one way so I felt I had to press the vet for clarity. Like all the vets I have worked with, she was compassionate towards wild animals and prepared to do what she could, but also realistic. As I suspected, she was being delicately euphemistic, and I continued to hold the scrawny creature, laid down once more, while the lethal dose was delivered. She was so emaciated that I carried her to the back of the car quite easily on my own.

As well as vets, I often work with the police. Their presence is always useful, sometimes invaluable. One Sunday morning an officer I knew quite well called to say that she and her colleague had come across a deceased casualty and would wait on my arrival. The deer was a big winter-grey doe, the rural road narrow and fast with a series of blind bends and cramped verges. Without the help of the two officers, I can't think how I might have managed to deal with the situation. As it was, they were in an unmarked car but at least wearing hi-vis. I pulled up behind them and switched on my hazards. The doe had been hauled up on to the narrow verge. She had been dragged under the wheels: her pelt bore multiple scuff marks and one hindleg was de-gloved down to the ankle, exposing the dark red musculature underneath. Corvids had plucked out beakfuls of hair. I decided our best option was to carry her back up the road a short distance to where she could be lifted into a field above the head-high retaining wall. We were in a dangerous spot and needed to act fast. The woman officer stopped the traffic while her male partner in anti-crime and I picked up the heavy load. Without thinking, I grabbed hold of the hindlegs leaving my assistant with the bloody head end.

We lifted her up on to the edge of the field, then, finding a toehold in the wall, I hauled myself up after her. While the police officer wiped the blood from his tunic with a fistful of kitchen roll, I dragged the doe by the hind feet into a patch of scrub where the corvids could feast in safety.

One moment I'm sitting on the sofa wrangling over the killer Sudoku in the paper on a lazy Sunday morning, the next I'm in the car driving impatiently, my mind in a tunnel, unthinking. Soon I've got hold of the back-end of a full-grown deer, I'm testing the gaps in the wall with my fingers and toes, I'm dragging a heavy carcass into cover while a police officer, spattered with blood, wipes himself down. It's all over in a flash; a compressed sequence of short bursts of exertion laced with excitement and purpose. It seems to me I am most alive during these brief flurries. I try and picture another, still more concentrated sequence of events, even more crammed with life and energy: the accident from the driver's perspective; and from the deer's.

For the rest of the day I'm buzzing. More than anything else, my life is now measured out by these rather unconventional episodes, which are like knots on a length of string or blips on the flat line of an oscilloscope. In previous years there were so many badgers and the incidents were so alike, almost routine, that it is difficult to single many out in retrospect, but each DVC is quite distinct in my mind, with its own quirks and individual qualities. I don't crave their happening, I don't wish them to take place at all, yet the excitement they engender is undeniable so that my sense of purpose is tinged with guilt. In between times, like a soldier starved of action, I may admit to being a little bored.

It seems an unfortunate and shameful paradox to arrive at, but on a day when a deer has died, I can say that I have lived. Yet any guilt might be assuaged if I remind myself that some-times – just sometimes – there has been a happy outcome, and not for the foxes and corvids for a change. There have been days when the deer lived too.

The first casualty that was saved was a yearling doe. She was taken to surgery and X-rayed. No bones were thought to be broken. She had suffered some trauma to the head, however, and one eye was closed and swollen. When we took her from the vets to the rescue centre, she was groggy with drugs and didn't seem quite real, more like a clever animatronic model whose finer adjustments were a work in progress. Her head was up but unsteady, her legs wouldn't work. The hold-ing pen at the centre had first to be emptied of ducklings and refreshed with dry bedding. She was laid on the floor of the shed beneath a heat lamp. We closed the doors and turned out the auxiliary lights. This was the critical period. I didn't expect her to last the night.

In the morning, however, I was pleasantly surprised to find the deer on her feet. She was still dopey and barely protested when her manky eye was bathed and various cuts and grazes were salved. She was given a bucket of water, a bowl of ewe's milk (she was only young), 'sheep mix' pellets and a tray of fresh-cut greenery which I had gathered. Then we left her alone.

On the second morning she seemed a little brighter and showed some curiosity towards the world beyond the shed when the door was opened. But she had shown no interest in the food options provided, only being observed to drink from the black plastic bucket while a dainty hoofed foot stood in the bowl of milk. Nevertheless, there were a handful of shiny black droppings on the floor which, far from arousing disgust, felt like a payout from a slot machine. We gave her fresh milk in a second bucket, identical to the first, and tried wiping milk around her mouth, but there was no encouraging response. The damaged eye was a mess of raw flesh and pus but she was still docile enough to let it be treated.

A vet came to look at the deer in the afternoon and advised that the eye would take several days to heal and we shouldn't think of releasing her until then. In the evening the head of staff at the centre rang and excitedly informed me that they had succeeded in getting the young doe to eat, having force-fed her with a mush of dried food intended for rabbits softened with warm water and loaded into a customised syringe.

A book about caring for wild animals recommended ivy and rose petals to tempt a captive deer's appetite. It also stated that a deer that refuses to eat *voluntarily* is unlikely to recover.

Ivy was easy enough to procure and I also bought a sack of moist 'goat mix' from a livestock food wholesaler. By the third day the deer had taken to lying down. She looked more comfortable like that but we weren't sure whether it was a good sign or not.

It was early spring and traces of overnight snow were apparent on the fourth morning. Even without the rose petals, the doe's untouched food choices had assumed the air of offerings presented to a graven image. Nonetheless, the living animal was becoming difficult to restrain, which had its positive and negative sides. I held her underneath while she was being treated and force-fed and could feel her heart beating strongly. The bad eye was healing and her good one seemed to grow larger, rounder and darker by the day as she slowly but surely improved. You could fall right into a deer's eye and, floating free, wash up on a far shore. The main problem when holding a lively deer casualty is keeping the head still as it waves about on the end of a long, muscular neck. When I finally let her go, my hands were covered in a black grease and stuck with loose hairs.

Then there was real progress. When I peeked in on her the next day, she was actually munching, head down, on the wreath of ivy in the corner. On the sixth day, very early in the morning, we released her close to where she had been found, but well away from the road.

The next deer that was saved was more than two years coming. There was some uncertainty as to what had happened. The buck was found lying just inside a rickety gate by the roadside. His limbs were sound and he was breathing normally, but he couldn't get up or remain standing if helped to his feet, and he kept his eyes closed as though wishing himself somewhere else. The farm and equine vet thought he might be concussed and gave him painkilling and anti-inflammatory injections. Either he had been struck a glancing blow by a vehicle or he had had a tussle with the gate in trying to get through it. The vet, a young woman of unremarkable stature but remarkable strength, insisted on carrying him by herself to a safe place I had identified in the middle of the fallow field where there was a group of trees and dense undergrowth. I said I would check on him in a couple of hours. The vet gave me a syringe and recommended that I try and get some water down his neck if he was still alive.

When I returned I could see that the buck's head was up and he was holding a natural resting posture. I phoned my original caller since, if his strength was returning, I would need some help restraining him. She brought a friend. The friend held the buck down gently but firmly, I held on to his rough antlers with gloved hands and the other administered the water. The buck made a vain attempt to get up but his eyes remained closed. He responded to the water with an automatic chewing-sucking action and we took turns at gently massaging his throat. A little blood had come from his nose. He was a mature individual in prime breeding condition and potentially dangerous. When we let him go, his head drooped and swayed on his long neck as if he were drunk. The two women were enchanted with him and I had some difficulty persuading them to come away. It was drizzly so I draped a towel over his body. Once again, he looked unreal, like something out of a fairy tale. Perhaps he would disappoint us by turning into a prince.

In the morning I found the towel discarded on the ground like a sloughed-off skin. It was damp and dirty. A strange sound was coming from behind a screen of brambles. He was evidently in there and I could hear him breathing in an exaggerated manner which suggested fear or rage, or both. The buck wouldn't be happy to see me; he would associate me and others like me with his torment. But he was apparently himself again and that was more than I could have hoped for. The combination of things we had done for him seemed to have done the trick.

I thought I had saved another earlier in the year. This deer, also a buck, had been caught in a fence and someone had cut him free. But now he was 'stuck' somehow in water.

I found the buck standing in the splash pool below a high man-made waterfall. He was visible from the road bridge and a number of spectators had been keeping an eye on him in shifts for a couple of hours or more. I knew straight away that I had to get him out. It seemed to me that the deer must have been disorientated after its initial ordeal and had staggered away up the river – I had come across deer using rivers as safe corridors before – until it came to this insurmountable barrier. The people on the bridge only added to its confusion and, I concluded, it felt trapped. Standing in cold water all this time wouldn't have done it any good.

I straddled a couple of barbed wire fences and lowered myself down a vertical drop on to slippery stones to get to the head of the fall. When I got the buck's attention, however, all he did was lower himself into the water as if trying to hide, and get wetter. I needed to get closer. So I approached from downstream this time, removing my boots and socks and rolling up my trousers. I edged around the bottom of the fall on a treacherous ledge beneath the cascading water until I was standing right next to the buck. Still he didn't want to move. There was string wound inextricably round his antlers. I asked an onlooker to pass me down a stick, but when I attempted to push the deer with it, it broke. I called for a better one and was handed a stout upgrade. The buck turned his head and looked at me pathetically while I prodded him in the rear, but eventually I managed to push him in to where the water was deepest and he could swim in the direction I wanted him to go. Under normal circumstances roe deer are strong swimmers. Passing beneath the bridge where a few people were whooping and clapping, he climbed out of the water on to cobbles and began to nibble at a long beard of ivy hanging down. But I wanted to keep him moving towards safer ground so I encouraged him onward. However, he lost his footing and fell, and I realised then how exhausted he must be. Fortunately, there was a strip of turf below the high retaining wall and the late afternoon sun was shining down on it. Regaining his feet, the buck was drawn to this little patch of Eden. He lay down. Now I thought it best to leave him and hoped he would recover his strength and nerve. I would come back after a while.

Two and a half hours later I looked over the wall from the road. The buck was lying, unnaturally, on his side. He was dead. I called a friend to give me a hand in moving him. His tribulations and the associated trauma had been too much for him. Deer are such sensitive, otherworldly creatures and prolonged exposure to human contact may kill them.

12

Winterlude

WINTER IS A quiet time. Few casualties are reported around the turn of the year. Badgers, the most frequent large road victims by far, are less active, sticking close to their setts, even neglecting to visit nearby artificial feeding stations which had previously sustained them. Most have reached their target weights and a hormonal switch has deactivated their voracious appetites. Underground, members of a clan at peace in the main sett feel the benefit of shared body heat.

When there is snow on the ground, badger prints are concentrated around the immediate area of the sett, as are their dung-pits, which, at other times, might be some distance away, perhaps indicating a territorial boundary, often in parallel with a linear man-made feature such as a fence, wall, hedgerow or footpath. Fox tracks, on the other hand, wander far and wide. Foxes must still find food to eat while badgers can live off their fat reserves. Winter is the mating season for foxes and, typically, a dog and vixen couple will stay close together on their travels. Around the turn of the year, there may be more fox than badger road casualties.

During one particularly cold spell, the snow lay on my field for so long it dried out and took on the texture of sand. Despite the covering of snow, conditions were dry enough for the badgers to gather dead grass for fresh bedding and their harvesting activities created chutes leading to and from the sett. I observed another unusual effect on the moors. My RSPB colleagues and I were planting trees in deep snow. The conditions were far from ideal but if we didn't get the young trees in the ground soon their roots would freeze and die. On our way out, crossing an exposed ridge, we came upon a set of mountain hare tracks which instead of being impressed in the snow were raised above it as if somehow the surface of the world had been turned inside out. We came up with a theory. The hare had loped along the ridge and left a trail of normal prints. The snow in the bottom of the prints was compacted and made firm by the hare's weight, but the raw wind blew the looser surrounding snow away and the exposed compacted snow in the shape of inverted hare prints froze hard and endured.

The coldest months are a good time for birdwatchers and, being at something of a loose end, I am apt to join in. The river is too high and fast and turbid too often for the king-fishers now and they seek out sunken moorland drains and millponds to sustain them instead. Cormorants and goosander settle on the reservoirs to the consternation of the human fishers. More sought-after waterfowl include the drake golden-eye with its swollen bottle-green head and big white facial spot like a fat snowflake. Or, if you're really lucky, a great northern diver like the one I saw taking crayfish at the sailing club. The head and neck are pulled back like a handle or lever, there's a slight lurch forward and the great goose of a bird slips under the surface of the water. There are birds of note almost anywhere. Occasional winters there are irrup-tions of exotic waxwings, their flocks descending on crops of berries in residential areas: in supermarket car parks, school grounds and on tree-lined streets. From the damp woodland floor I flush woodcock, pieces of the mind-boggling puzzle of leaf litter spirited into flight. Once I was handed a dead snipe, a victim of the big freeze of that year, the tip of its ridiculously long beak flexible and riddled with sensory pits for detecting subterranean prey at some remove from the bird's more familiar sense organs. But a long, touch-sensitive beak is of no use if the ground is frozen solid.

The average twenty-first-century birdwatcher can track sightings on social media and follow up on those which are of particular interest to them. Stuck in a mock-heroic past, I take the old pack-horse route that winds through bleak peat bog out to a remote wind-tossed, brim-full reservoir. Perhaps there will be a lone winter visitor I can have all to myself. But there is not a living thing to be seen or heard. Not so much as a grouse on the barren moor or a Canada goose on the dark, choppy water; not even a crow in the whole expanse of grey sky. Only a metaphysical poet could find inspiration in such utter blankness.

Later in the year which saw the return of the sow badger Bertha, I organised the introduction of a second sow in a wood just two miles from my flat. She was from out of the area, had been dehydrated when rescued, and was kept captive through the long dry summer until she had recovered her strength on a diet of day-old chicks and peanuts (apparently she wouldn't touch pet food). There were no resident badgers in the wood and we had installed an artificial sett there a couple of years earlier which still hadn't been taken up naturally. However, there was evidence of badger traffic passing through so it was hoped that, introduced to the man-made den, she might attract a mate and raise a family. Since I was to take sole charge of her daily needs, I renamed her to my own satisfaction, calling her Bonny after the name of the wood.

I kept up the provision of day-old chicks until the supply in my freezer ran out and then, by the time she was free to roam, I limited her to nuts and dried fruit, hoping to encourage her to forage of her own accord, which she did with gusto, turning up the rich humus over a wide area. I recognised her dung-pits outside the gate to the wood from pieces of undigested fruit pit and nutshell in her otherwise loose muddy motions. I was pleased with Bonny's progress but worried that if she was, after all, to spend the winter alone in a chamber made out of breeze blocks, she wouldn't be warm enough. So I supplied her with armfuls of dry straw and hay, which she took in backwards down the long tunnel of plastic pipes to her sleeping quarters.

A dead tree – standing dead wood – must have been under-mined when the sett was excavated by mechanical digger, and one night it slammed down on top of the sett, but perhaps Bonny was safely out and about at the time of the fall since she hadn't been scared away. I kept up my daily visits for well over a hundred days. When the snow was too thick or the ice too slippy to get my car out of the communal car park at the mill, I trudged up the hill and continued my provision. I could see Bonny's tracks in the snow where she had emerged from one pipe, turned over the stones to access her food and then gone back in at another entrance.

One day towards the end of winter, as I approached the wood as usual from above and could see all the way down the hill to the small rectangular reservoir at the bottom, I noticed there were some large white birds on the water. In great excitement, I set off down the scrubby, muddy hillside by the most direct route to get a better look. I had been right, they were whooper swans, ten of them, the most majestic of our winter visitors. These straight-necked wild swans with yellow-and-black bills like sharpened pencil points are transi-tory apparitions at best, rarely remaining in one place for long, at least in these parts.

The swans drifted idly this way and that, the odd bird exercising its great wings while standing on the surface of the water, another whooping in a half-hearted, querulous manner. I hung around long enough to witness them take off again, after an hour or so. A ripple of excitement passed through their number – they were all whooping at once as though voicing agreement – and they rose in formation, joyously, and performed a couple of triumphant-seeming flypasts before heading off in the decided-upon direction.

I also saw green woodpeckers and deer on some of my regular visits to the wood, but all this time I never actually saw Bonny. It was as though we were engaged in a form of correspondence, except the badger's contributions were unintentional. Sometimes I obtained photographic evidence of her continued occupancy. By late spring, however, the correspondence was at an end and the sett appeared to be empty. I like to think that Bonny had been accepted into a nearby clan.

A year later the sett would be breached. During the interim period I had checked it for activity on a regular basis while monitoring others in the area, but the only tenants I would swear by were rabbits. The diggers went to a lot of trouble. They located the chamber and dug a trench, exposing the protective layers of weld-mesh below the surface. The top layer they bent back, then they cut a square hole through three more layers and inserted their dogs. Tufts of fine hair of two different shades were caught on the sharp ends of the wire.

Bonny had excavated her own narrow exit hole above the chamber and this, too, had been enlarged by the terrier men. It was highly probable the perpetrators were the very same who, several years earlier, had dug the nearest natural main sett on the edge of a steep ravine. In both cases, little attempt had been made to hide the evidence of criminal damage. I could only think that they had expected to secure a badger on this occasion, but was confident there were none to be had.

Yet, there was a lesson to be learnt for badger protectors. A so-called 'protected' sett was not invulnerable after all.

Returning from the wood one morning, I fell into conversation with a woman walking a small, timid dog. She told me something which made me rather envious. The previous year she had been attacked by a deer. Not a buck, fortunately, but a doe protecting its kids. She had entered the meadow below the wood unwarily with the dog on its lead when the doe came boldly towards her and butted her low down on the leg. Instinctively, the woman, fearing for the little dog, had bent down and picked it up and turned away. Now the doe butted her again, forcefully this time, in the small of the back, pushing her up against a fence before the woman could escape. It seems clear to me that the dog was the intended target of the deer's wrath. A fox would have been subjected to similar treatment. I was also impressed that the woman bore her attacker no grudge, but understood and accepted its behaviour.

Most days, I had the wood to myself. Sometimes it was held under a spell. Only I could move through the trees and my movements were dream movements. Everything else had been turned to frost. Each twig grew ice spicules like crystal thorns, and as the grip of the frost tightened and the dream swelled, the thorns branched, thorns on thorns, like a fractal. If a single bird was to flutter and fly, the whole icescape would shatter and crumble into dust.

There was a severe outbreak of avian flu that winter that manifested itself mainly in seabirds, waterfowl and birds of prey. On New Year's Day, friends and I paid a traditional visit to a usually productive location, a country park with two ornamental lakes, where we could get our bird year lists off to a good start. But we ended up counting dead birds floating in the water: heron, cormorant, Canada goose, greylag goose, mute swan – the big, majestic species seemed to be most susceptible (or conspicuous). There were no goosander to be seen, which was hard to believe. One doomed goose, unable to hold up its head, was spinning in an eddy.

Animals become tamer, bolder, more desperate when they are cold and hungry. Like the robin which would fly to my hand for seeds when I was house-sitting for friends. Or, during the same sojourn, the bank vole which scrounged at the foot of the bird table each day. A colony of rats thrived beneath the bottom shed in my compound during one winter. Like a Jain monk, I gave them their own bowl of food, although they were perfectly capable of accessing the bird feeders squirrel-fashion. The rats' excavations extended under the perimeter fence and once they had abandoned the warren they had created in spring, the hole under the fence was enlarged by bigger animals. Now foxes and badgers could get into the compound, which was something I was happy to go along with.

Hibernation is a less ambiguous strategy for survival. There is a derelict farmhouse in this valley, its roof now fallen in and a mature tree growing in the doorway, which nevertheless retains three discrete rubble-strewn wine cellars with vaulted ceilings that make me think of the three underground chambers in 'The Tinder-Box', with their increasingly valuable hoards of treasure. In the driest of the three, and the least invaded by tree roots, I can find hibernators by torchlight: a handful of small tortoiseshell butterflies, perhaps, and a score of herald moths suspended from the whitewashed bricks overhead where they are safe from mice. Only a few insects hibernate as adults; a few more butterflies (peacock, comma, brimstone), ladybirds, queen wasps and queen bumblebees. Some make the mistake of entering our centrally heated homes only to be repeatedly aroused from sleep, which requires energy, and they ultimately perish.

When I am occasionally called out during the late or early months of the year, there is a good chance it will be after dark. This circumstance adds to the disagreeableness of the task under the prevailing weather conditions.

At the end of my first year as the Deer Man, there had been exactly fifty casualties that I knew of, at least forty-four of which were the result of road accidents. One was hit by a train, two died after being caught in fences, one was killed by an Alsatian and the cause of death of the remaining two was not known. Of the forty-four road casualties, only one was a kid; the gender split of the adults was roughly half and half. Fifteen (more than a third) were still alive after the collision. Of these, ten were euthanised, one died later, three got away and one recovered after treatment. Twenty-four (more than half) of the road accidents occurred in the three-month period April to June. This coincides with the period when yearlings are ousted by their expectant mothers and become independent for the first time, making them particularly vulnerable. A doe killed in June had been lactating; I was unable to locate her offspring.

A few of the corpses had vanished by the time I arrived on the scene. I have learnt that deer carcasses are sometimes claimed by chance-comers for human or canine consumption. But I was able, sometimes with help, to relocate twenty-three of the carcasses to where they were unlikely to be found by people, never mind taken away for personal use. If I hadn't done this, the chances are that the corpses would have been incinerated by professionals and gone to waste. In many cases I followed the course of decomposition or natural destruction over days or weeks.

13

New Year

I WON'T PRETEND THAT my fellow badger buffs and I drifted apart. It was much more drastic than that: a rift opened up between us like a crack in the earth's surface, unexpectedly, and at first I was swallowed whole. It was dark down there, but I clung to the belief that one day, under my own volition, I would emerge back into the light.

The people I relied on for help when I most needed it were no longer there. Each deer casualty, dead or alive, from now on would be a potential challenge.

The first casualty of the new year – on the seventh day – was a live one. It was a Sunday night and I knew that getting a vet to come out in emergency hours would be a problem. No one had stayed with the stricken animal. However, I had an incredible stroke of luck, as the next car after mine on the lonely, unlit road pulled over and its driver, who had spotted the deer, its head up and big eyes catching the light, quite by chance, happened to be a vet on her way home with her partner.

The casualty, a young doe, was quite lively, but an injury to its left hindleg prevented it from jumping over the high fence on either hand. Instead, it got up and half ran, half stumbled, along the top of the embankment and we were able to close in and pin it down. I took a firm hold of the doe's forequarters while the vet checked her over. The doe struggled gamely and emitted strangulated cries, but she was going nowhere. The vet thought the injured leg was unbroken, that the deer had probably 'taken a good whack' and was badly bruised. She decided that the best thing would be to get it into a field and away from the road where it could settle and she would check on it again in the morning.

Where the fence ended on the opposite side of the road, there was a drop down to a lower field. It took the three of us to manage the manoeuvre, two of us carrying at first and then lowering the deer down to the third person in the field. On its feet, the doe staggered away from us once more, then lay – half fell – down, but we moved her on again until she was over the brow where she wouldn't be disturbed by passing headlights.

There was no sign of the deer the next day. The vet and her partner, who lived nearby, had spent half an hour searching with torches while it was still dark, and I looked once it was light. It was bitterly cold and the scene looked different by day so that I had to prove to myself I was in the right place. There was a tumbledown fence at the bottom of the field we had released the doe in and a patch of scrub beyond that, then another minor road, more fields and a wooded clough, but taut wire fences criss-crossed the landscape and excluded the lame. That a casualty given the okay by a vet should get to its feet and take itself off is as good a result as we can reasonably hope for, but question marks still remain about its chances in the longer run.

A deer–vehicle collision may be a useful metaphor for a world spinning out of control and its impact on a fragile, barely considered, older and ultimately more precious world.

We refer to deer as a driving hazard, shifting the blame and responsibility on to the animal. The damage done, but exonerated, we drive on again, only able, by artificial means, to see a short distance into the darkness that lies ahead.

Towards the end of the month the weather turned mild. The wind had dropped and the rain had stopped. A few birds were singing, specifically song thrushes and dunnocks. A friend and I set out for the moors. It was a day off for Sean from his busy work schedule and he appreciated the total absence of other people. Right up on the exposed heather-clad tops, we flushed a couple of deer from a gulley. They bounded with ease over terrain which caused us to stumble at every other step. Their targets were splayed – a sign of alarm – and dazzlingly white. Their dun winter coats matched the drab vegetation so well that their striking rump patches appeared disembodied, like pure white birds floating low above the ground. Through my binoculars, I could see that they were both does. I was surprised to see them up here at this season, among the grouse and heather. Perhaps, like us, they were hoping to avoid humankind.

On our way back down in the afternoon, we had rejoined a farm track when we spotted a fresh sparrowhawk kill temporarily abandoned up ahead, a woodpigeon with a bald rump and surrounded by an explosion of plucked feathers. We must have just interrupted the act of predation.

'It's breathing!' Sean exclaimed with horror.

I picked the half-dead bird up and Sean turned away while it fluttered and I wrung its neck. I left the corpse there for the raptor to finish and washed my bloody hands in a puddle.

My phone was ringing. A deer had been attacked by a dog and was now in someone's back garden. It was turning into an eventful day. Sean needed to get back so I dropped him off at the next village, from where he could catch a bus home, and drove on.

Personal knowledge of any wild creature, no matter how familiar, is a mutable as well as a growing organism. This latest victim was about to defy one of my previous claims (making me an unreliable narrator).

The deer had appeared on a municipal nature reserve and been pursued by the dog down a ginnel and into the road. The distraught owner had reclaimed his dog while a helpful bystander opened a gate into a backyard where the panicked deer took refuge. The homeowner had now returned to find the deer in situ and her car was parked in front of the closed gate. She was kind enough to invite me in with my boots on, to view the deer through the window without distressing it further. By now, a number of people had informed me that the deer was badly injured and bleeding, but it was mostly Chinese whispers. Said animal, a buck in velvet, stood in the concrete yard looking back at me. One hindleg was held clear of the ground. There was something odd about the leg but I couldn't tell what until the deer eventually put it down. Now the deer's posture was hunched like that of a hare. The leg was missing a foot. When the deer moved, it shambled, but its deformity was the result of a historical injury, completely healed, and the buck had clearly been getting by. It could walk, after a fashion, on the stump.

The way to age a roe buck is by comparing new antler growth. Mature bucks regrow their antlers first so that they can take possession of the best territories. This unusual buck was a rather small specimen but his antlers were well advanced for the time of year. He was only the size of a yearling but his 'head' was that of an adult.

Someone had called the police. Two baby-faced officers were on the doorstep. They had alerted the firearms unit and a veterinary practice. They looked at me dubiously when I told them I was the Deer Man. I protested that the deer didn't need euthanising. It had received no visible injury from its latest ordeal and, despite its unusual disability, it seemed to be perfectly chipper. I admitted that I had never seen anything like it but was confident that it would be okay and adamant that it deserved to live another day. I had won the police officers over and when I told them of my plan, they phoned for reinforcements.

Now the vet, a stethoscope dangling from his neck, had arrived. His was a face I didn't know. He too looked at me with suspicion when I explained who I was and might, but for a measure of decorum, have guffawed in my face. I had no official status in my new role; I was wholly independent and self-styled, and relied on authority figures knowing who I was from experience. The vet was appalled at the deer's injury, and I felt certain that if I hadn't been there to fight the buck's corner it would have been doomed. He referred to the animal as a 'prey species'; I don't know what predators he had in mind. I put my foot down; I had worked with many of his colleagues and I knew deer far better than anyone else present. And I won out. The homeowner declared she wouldn't have the poor creature put down on her property, and that may have tipped the balance. Once he knew of my plan, the vet took charge and acted as though it had been his idea.

Quite simply, we needed to get the deer back across the road safely and, as long as we formed two cordons a few yards apart, the deer would be channelled in the right direction and would know where to go. Why didn't we carry it across the road? an officer wanted to know. Because, I explained, that would involve unnecessary trauma and risk of injury to the buck's sensitive half-developed antlers.

There was a problem, however. The house in question was next door to a nursery school and cars were arriving to pick up small children. There was a general bustle – cars were blocking the path the deer needed to take and the animal was unnerved by the hubbub of voices and other attendant sounds.

Fortunately, the baby-faced cavalry had made it and the police took the situation in hand. The road was cleared and blocked above and below with stationary vehicles; one of the young officers backed the homeowner's car out of the drive and wedged it among the other parked vehicles. We all spread out in two lines: police officers in hi-vis stab vests, bystanders and myself. I handed out a few towels which I keep in my car, to be held out like matadors' capes. The homeowner stood on her front step ready to record the spectacle on her phone, and the vet volunteered to open the gate and flush the deer out of the yard after he had rallied the troops with the assurance, 'It's going to be more afraid of you than you are of it.' Everybody else was to make themselves big, but I took up a position on the pavement next to the ginnel where there was a gap after the last vehicle – and I ducked down so that the deer wouldn't be able to see me until the last moment.

Everything went according to plan – my plan. Although the deer had hobbled awkwardly in the confined space of the yard, it now set off at a lumbering, but serviceable, three-legged run down the drive and across the road. When it got to my side, it paused and looked at the gap at the end of the cordon, but now I popped up with a towel spread wide, and even though the ginnel began with a flight of steps, the startled animal cleared them almost effortlessly and shot up the passage towards the open ground higher up. I pumped the air with my fist and the homeowner was thoroughly pleased both with the outcome and with her video clip, which would go on social media.

I have seen and heard plenty of evidence of three-legged foxes coping well enough in urban and suburban settings where scavenging is a reliable alternative to hunting, but this was a first for deer. I looked closely at some photographs I had taken when the deer had been still, and saw that almost all of the lower leg was missing from below the mid-joint, which is actually the ankle rather than the knee. The stump was darkened and presumably hardened. A road accident or fence entanglement were the likeliest causes, but I usually find that similar wounds fester and the victim progressively weakens and dies. This chap was a true survivor and deserving of respect.

I phoned Sean to let him know what he had missed. He isn't such a wuss. On a previous outing he was bold enough to release a crow from a Larsen (funnel) trap which we spotted some distance away on upland sheep pasture. While I stayed with our rucksacks on the footpath, he used a drystone wall for cover then climbed over when he was opposite the trap. Commercial traps like this one are about the size of a budgie cage, but a home-made version hidden in a shelter belt may be as big as a walk-in aviary and lure in several corvids at a time with a sheep carcass for bait. A crow in a trap, like a deer chased by a dog, is a victim of circumstance, and some of us see it as our duty to give the unfortunate creature its life back when and where possible.

My old friend Laurence was back in the country having recently obtained Spanish residency, and he had been joined by his French partner. Before meeting up with them, I did my homework and discovered that roe deer in French is *chevreuil*. Marion informed me it is pronounced 'shevroy'. Like the Latin *Capreolus*, it is a diminutive of the word for 'goat', which explains why, to purists like myself, the fawn is a 'kid'. Marion's passion is foraging for and cooking with wild fungi, but she responds enthusiastically to all things outdoors and natural. I showed them the badger (*blaireau*) sett on my land, which was starting to show signs of renewed activity, of early spring cleaning. Last time they visited, it was summer and a big boar badger had been killed on my road (the one which bypasses the mill). Now I showed them its immaculately preserved skull, every tooth present and correct, and pointed out the features which make it so resilient. When we went for a walk through a wood, I would draw their attention to any deer sign – hoofprints or nibbled vegetation – and, with three fingers raised on either side of my head, intone 'shevroy' in my best approximation of a French accent.

I took them to a place where, the week before, I had seen a group of five deer grazing out in the open in the middle of the day. The best way to tell the sexes apart when they are head down and the bucks' budding antlers are hard to distinguish at any distance, is to focus on the rear-ends. Only roe does have a tush, the tuft of creamy yellow hairs hanging down below the stark white rump patch. Since it is inert, it is not a tail, but may act as a sort of flag when a mother has young kids to lead. This group appeared to consist entirely of yearlings: two male and three female (in the following weeks I would see additional, mature, group members). At dusk the previous autumn, I had seen three barn owls in this same location, taking it in turns to perch and quarter.

Normally speaking, the deer should be crepuscular too, but they are safe and undisturbed here; there is no public access through these rough fallow fields free of livestock. It may also be that they needed to feed for longer while the greenery was nutrient-poor. Disappointingly, there was nothing for my companions to see on this occasion; I could only tell them what they were missing.

The middle of February was absurdly – some might say, alarmingly – mild. One lunchtime, I got a call from John, who was at work. He had third-hand information that an otter had been killed on the road outside a well-known tourist attraction. 'I'm on my way,' I assured him. John lives on the eastern edge of the borough, whereas I am on the western edge. He has obtained first-rate trail camera footage of otters on his patch (of different individuals visiting a sprainting stone under a bridge). This casualty was right in the far north-eastern corner. It would take me half an hour to get there, if I was lucky. I have seldom driven with such impatience.

The otter, an adult female, was in the middle of the A-road. It wasn't in great shape. Hundreds – probably thousands – of busy people had driven by one of the most fabulous things they will ever see and no one had reported it to an appropriate body, or if they had, no one had responded. We know that at least one person, a birdwatcher/lorry driver, had stopped and taken a photograph – the first link in the chain which led to me (the fourth link) – *but had neglected to move the precious find out of the road*. My dream of possessing its skull – which I might have shared like a joint sporting award with John, taking it in turns to gloat over and show off – was shattered. Its head was more or less squashed; some of its entrails were protruding. But there was still much to marvel at: the long, laterally flattened tail; the webbed toes; the luxurious pelt; the long, stiff whiskers.

No sooner had I set off with my battered prize than I pulled over again. There was a rabbit hopping about in the road. It had myxomatosis, its eyes were puffy and sore; it was probably blind. A line of traffic built up as I gathered the pathetic creature up in my bare hands. I could feel it jump in its skin with surprise when I picked it up. This time I refrained from dealing the death blow and put it over a low wall. I can use the evasive excuse that I felt hurried by the press of onlookers who wouldn't see what the specific issue was.

In the evening I drove over to John's with my precious cargo. We weighed and measured it and speculated as to its movements. She had been a long way – miles – from any river, but another interested party (the second link) sussed out that there was a stream nearby leading to a carp pond. It seems likely that she had been there or was on her way.

I often refer to my familiar beasts as the Big Three: badger, fox and roe deer. But this otter was the Big One, the one I had been waiting for for so long. For me, the magic of otters dates back to childhood and a family trip to the local cinema to see *Ring of Bright Water*, to youth and reading Gavin Maxwell's wonderful books; to young manhood and pilgrimages to the west coast of Scotland to visit Maxwell's resting place and spiritual home. Here I will never get to know otters like I know the Big Three, but I can at least carry on dreaming.

The next day was miserably wet (again). I took the she-otter up to my compound and blocked the hole under the fence so that a fox wouldn't be able to carry her away. The previous day's excitement had bubbled over and subsided. Lying in the middle of the road, compared with the occasional weasels and stoats – and, once, a polecat – I had encountered, she had seemed so big and impressive; now she looked bedraggled and vulnerable, and the rain which was falling insisted on the sadness of the moment. All her admirable physical adaptations were meant for life – a singular existence – not this ignominious end. I laid her out carefully beneath a tree. Beginning at this time of year, the process of decomposition would be drawn out.

The extraordinary start to the year continued. The next week brought another first. I responded to a deer casualty on behalf of the council in the usual way. Just occasionally, a deer reported through this channel turns out to be a sheep, and when I set eyes on this beast lying on its back in the undergrowth, pale swollen belly uppermost, my first thought was that it was a livestock animal. It was certainly too big to be a roe deer; twice as big; the size of a bullock. Then I remembered hearing rumours of ghostly inhabitants of the nearby wood in the south-east corner of the borough, and looked more carefully at certain physical features. It was a fallow buck with knobby antler buds in the shape of little minarets. The corpse was partly scavenged, the upper flank violated and the ribcage peeping through dark muscle. Blood diluted with rainwater had pooled in the cavity. My enquiries back-dated the death to three days earlier. I can drag a roe carcass or carry it a short distance on my own, but this beast was in a different league. Fortunately, it was on John's patch.

I arranged to meet John at the scene before dark but I got stuck in traffic. He is a builder and much stronger than I am, and by the time I arrived he had dragged the carcass across the road and stashed it under brambles. It was leaking watery blood so profusely that he abandoned the idea of getting it into the back of his estate car. Now that I looked again, I saw that both of the buck's forelegs were broken. The setting was a 'quiet' lane, the sort of semi-rural back road where, despite the lack of a pavement, locals walked their dogs in safety. For a large, out of the ordinary animal to meet with such a violent end in these circumstances seemed obscene, outrageous.

Car headlights flay the dark like scalpel blades and expose phantoms.

One of Hans Christian Andersen's late stories is entitled 'The Most Incredible Thing'. In a sort of talent contest, whoever could present the most incredible thing to the assembled court would claim the hand of the king's daughter in marriage and half the kingdom. A thing of wondrous human artifice is defeated by an act of vandalism, but there is a happy ending after all.

For me, the most incredible thing during the first quarter of 2024 wasn't the three-legged roe buck or the road-killed otter or fallow buck. It was the red stag with the bizarre appendage. When I finally saw the stag in the last week of March, it was the first I had heard of him in weeks, for I had first got wind of his plight over two months earlier, but there had, it seemed, been nothing I could do at the time.

I received snippets of information. The location was slightly out of area but not far from my address and a place with which I had some familiarity, having visited the site of a one-time mink farm, its long barn still standing, a handful of times. The deer, said to be a red stag, had been caught in a fence, but police officers had cut it free. However, it was now carrying a detached fence post and the police had issued a warning to the public not to approach the animal. And they were right; here was a powerful, dangerous beast. Then I heard nothing more and imagined that the deer would, sooner or later, succumb to exhaustion.

So the much delayed next report took me by surprise. This time I went to investigate. It was late afternoon and, mercifully, it had stopped raining. The deer wasn't hard to find. It was indeed a red stag, a fine specimen – *he looked in excellent health.* A six-foot fence stake hung vertically from his left antler. When he lowered his head to graze, it made contact with the ground like a wooden leg. A tangle of cable was wound tight around both antlers, securing the heavy post. Used to his local celebrity status, the stag was out in the open, in the middle of a green field, pretty much unperturbed by my close scrutiny. My caller had informed me that it was in company with two 'baby ones'. I knew they would be roe, and there they were in the rough, a buck in velvet and a doe. To add to the spectacle, a barn owl glided by lower down the field.

While roe bucks are gaining new antlers, red stags are about to drop theirs. Any day now, this stag would be free of its torment. Not that it gave any impression of being tormented. It seemed to have adapted admirably to its unusual circumstances. Like the three-legged roe buck, it was a survivor. The stag was close to its prime and next rut might well secure for itself a harem of hinds, in order to pass on its survivor's genes to the next generation. In the meantime Easter was almost upon us and the encumbered stag's timely public appearance suggested a grotesque Passion Play.

On the fourth day, which happened to be Maundy Thursday, the stag was just as conspicuous by its absence. I looked for big slots in the mud on tracks and in open gate-ways on all sides, but could find none.

For the three days while the stag was in plain sight, there was nothing explicit for me to do. But it still felt important to be there, to observe the subject and talk to passers-by and other interested parties, to ascertain and confirm all the known and knowable facts. It was possible to put people's minds at rest. Yes, he had been like that for more than two months, but the shedding of his antlers was overdue. Outside the breeding season, it was not unusual for him to be solitary or a long way from the rutting ground. No, he hadn't escaped from the nearby deer park; they were sika. No, he shouldn't be darted because it could take fifteen minutes for the drug to take effect during which time . . . And, no, he shouldn't be put out of his misery since he wasn't miserable. He was hold-ing his head up with something like pride and perhaps contempt for the slight inconvenience he bore.

14

The Big Three

Throughout these years there have been half-acknowledged goals and dreams lurking at the back of my mind. One was that my land should provide sanctuary not just for badgers but for the Big Three, as I like to call them – that is, for foxes and roe deer as well – our three common larger mammals; three species which manage to thrive, despite the worst we can do, by virtue of their deeply ingrained survival instincts.

During the spring of this year, I was a little puzzled to find that my badgers – the Clough Clan – were active only at the lower end of their linear sett. That was where all the evidence of fresh digging was and where traces of recently gathered bedding were strewn. In previous years the cubs had been raised at the top end which, fortuitously, provided me with the best possible views of their early forays above ground. It wasn't until early June that I understood.

One late afternoon during a rare spell of fine weather, I did the rounds of my plot pretty much as per usual. As I walked along the top side of the dry-stone wall which delimits my field, I noticed a redness in the long grass just outside the sett, too red to be a mere concentration of sorrel. Two foxes were basking in the sun. One got up – it was a cub – ducked under the wire fence and slipped out of sight, but the vixen (long, sharply pointed face), with occasional fussy adjustments of her position, remained where she was for a good ten minutes. Then she suddenly became alarmed – she could only have detected my scent at last on the swirling breeze since there was nothing else to disturb her repose – and vanished like fluid down a drain.

Over the following days I was to learn that a litter of four fox cubs was occupying the top half of the sett, emerging from what I regarded as the main entrance, which was criss-crossed with tree roots, and another outside the fence. This second entrance had, back when livestock were kept in the field, been covered with a paving stone to prevent the beasts stepping into it and injuring themselves, but the slab had long since subsided and upended so that the hole was back in regular use. Now a little fox's head would pop up, as though through a trapdoor to the basement, to check that the coast was clear, before the cub hauled itself out.

My sightings of the fox cubs, sometimes with the vixen present, overlapped with my first views of the year's badger cubs. Perhaps the fox cubs' diurnal appearances had affected their badger neighbours, for I began to encounter the badger cubs by day too. In fact, these youngsters were more adventurous since they strayed well away from the safety of their burrows and foraged for food, unaccompanied by adult guardians, out in the open. There appeared to be three new additions to the Clough Clan this time around. Sometimes I was able to creep up very close to a cub, close enough to hear its long claws tearing at grass roots to turn over the soil for insects, before the busy little creature became aware of my nearness. Even then, it might not move very far away before resuming its industry.

It is well documented that fox cubs are sometimes raised in occupied badger setts, but up until that point all the fox litters I had watched had been born either in vacant badger setts or unequivocal fox earths. I have also read and heard that both carnivores are capable of killing the cubs of the other, but all seemed to be harmonious, at least on the surface, in the present case. I was as pleased as a dog with two tails.

Another goal or dream was that I might encounter a newborn roe deer. In the middle of June I had, one evening, gone to look for a doe that was a regular in a meadow across the road from my land. She was safe there since the fallow field was below the level of the road and, behind the belt of trees at the back, there was the railway line. Last year she had had two kids but I didn't see them until the middle of August when their spots had mostly faded and they were half their mother's size.

On this occasion I couldn't see the doe at first, but then I spotted her rump. She was bending down in the long grass, but I could tell it was her and not the buck because of her anal tush. She lifted her head a little and I saw that she was nuzzling a tiny infant. The kid was probably no more than two or three weeks old. As the doe left off nuzzling and stood at her full height, the kid came into clear view. Its flanks were warm brown with bright rows of white spots. The doe moved away slowly and the kid began to follow. It was already agile, jumping for necessity, rather than walking, through the long grass, which it occasionally mouthed. Its mother turned to nuzzle it some more.

I could hear someone coming along the pavement. It was a woman with a small dog. As she came near, I presumed she must notice how intently I was watching something and be curious to know what it was, but she was about to walk by. Keeping my feet still, I swivelled my upper body and caught her eye. She smiled. I probably looked harmless enough with my matching bush hat, cagoule and wellies, binoculars and day-pack rucksack. I put my finger to my lips and hissed, 'Look at this.'

The woman looked and gasped, and lifted up her phone which she produced automatically from somewhere about her person. The kid was momentarily hidden from view. 'She's got a little one,' I continued.

There was a stirring in the grass. 'Look, watch this.'

The woman gasped again as the kid reappeared, and carried on filming. The doe had noticed us by now but didn't seem unduly concerned that her secret was out. She began to drift away from us towards the back of the meadow, the kid continuing to follow in its own exaggerated fashion. The woman asked a few questions and I did my best to explain. I confessed to never having seen such a young kid before. We talked easily. 'You'll be glad I stopped you,' I said confidently.

'I would have walked straight past,' she admitted, and added, 'I'm buzzing.'

A little over a fortnight later I was called out to rescue a 'baby deer'. I am inured to reports of 'baby deer' casualties which prove to be adults and expected as much of this latest incident. The deer in question was apparently cornered behind the windowless back of a large building, a carpet showroom, in the dark channel at the bottom of a tangled embankment. My caller had obtained a photograph of the creature before it had hidden itself away. I asked to see it and was surprised to be shown a picture of a spotted kid.

My informant lived next door and had heard the young deer calling like a bird. She told me it had a bad eye. I fetched a pair of secateurs from my car and cut away a screen of brambles from the end of the dark tunnel. Then I strapped on my headtorch on its subdued red light setting and crawled in. A moment later I emerged with a month-old kid in my arms and, wrapping it in a towel, tucked in its spindly legs. She was alert but placid and quiet, which was a blessing. As with a small fox cub, she was short-faced. Her right profile was enchanting, but her left eye was a mess of pus.

I explained to my impromptu assistants that I needed a driver to get us to a vet while I held on to the precious bundle. I sat in the back of the car with my arms full while my driver drove with the utmost care and we were frequently overtaken. The kid was as good as gold and perhaps benefited from the warmth of close contact for she had looked bedraggled and exhausted. The vet came to the car – there was no question of entering the busy waiting-room with its canine clientele. Ideally, he would have liked to operate and 'enucleate' (remove) the infected eyeball, but I stressed the importance of returning the infant to its mother's care as soon as possible. Under the circumstances, we all did the best we could.

The vet administered painkilling and antibiotic injections and daubed antiseptic cream on the affected eye. When we returned to the place of the discovery, having neither hand free, I was forced to squirm my way up the overgrown embankment while the kid, lively all of a sudden, wriggled like an eel. It may have been the drugs or the familiarity of the surroundings, but she had begun to call out – which was a good sign.

At the top of the slope, I found myself confronted with a secret wilderness of meadow and scrub. Choosing a small tree that I might relocate in the morning, I put her down on her shivery limbs and backed away. I felt like I was abandoning her. The only hope now was the return of her mother.

Overnight, having had time to reflect, it occurred to me that if the kid's mother had two offspring, which is common, she may well have deserted this one, sensing that it was weak, to concentrate her attention on the other. This is one of the ways in which nature works.

When I looked again in the morning, I was rather surprised to find the kid nestled in the long grass at the top of the embankment close to where she had first got into trouble or drawn attention to herself. Had her mother suckled her in the interim? I really couldn't say. What was I to do for the best?

Innumerable times had I pictured this moment, of coming upon a spotted kid in the grass and the conflicting emotions it might arouse. The urge to take photographs, to touch, to possess; and the anxiety not to disturb, do harm or sully an image so perfect. Even though the situation was slightly skewed, it still felt like a significant high point in my life, a crowning achievement. Everything afterwards would be different somehow, coloured by the experience. Whether for better or worse, I couldn't tell.

I sought advice from the local wildlife rescue centre, which, as usual at this time, was full up. I was told that the infected eye was a killer and should be the priority. But the kid would need specialist care as well as treatment. Whitby Wildlife was my best option and, having been given the go-ahead over the phone, I fetched my carrying cage and returned for the invalid.

She had moved again, but only under a tree on the embankment. She didn't object greatly when I picked her up once more and carried her down the slope. The cage, fine for badgers and foxes, wasn't ideal since she couldn't quite stand up in it. I hoped she would settle among the towels I had laid inside. We had a three-hour drive ahead.

She was the ideal passenger since it was easy to imagine she listened with interest to my running commentary on the journey, much of which was familiar to me. After a long time she scuffled about occasionally, which I was grateful for since the noise indicated she was still alive. It would have been a grim return journey if she had been dead on arrival. As we came over the brow of the North York Moors, I announced the view of the sea. We were almost there.

The kid was taken from me and I was told that she would be tube fed directly into her stomach and examined shortly by a specialist vet. I felt positive about the trouble I had gone to and dearly hoped it had been worth it.

For technical reasons, I didn't receive news on the kid's progress for more than two weeks. Little One Eye, as I thought (sentimentally) of her, had proved too weak and unresponsive to intervention, and had been put down. That sultry evening, as if to compensate in part for this blow, I saw the mother – the regular doe I have previously referred to – together with her kid for the first time in five weeks. The kid would be almost two months old now. It attempted to suckle, but its mother stepped aside. I watched them for several minutes. Mum's ears flicked almost incessantly and she shook her head. If flies were the reason, Junior was relatively un-affected. The tips of the kid's ears reached its mother's shoulder. It had grown considerably and there was no longer any need for it to jump to progress through the long grass.

Acknowledgements

The evolution of this book has been a long and winding road as well as a bumpy ride. Whether the end result justifies the means is not for me to say. The car broke down several times. New parts were added and faulty ones scrapped. On this frequently interrupted journey, was I driver, navigator, mechanic or merely passenger (back-seat driver)? I'm not entirely sure, but what is certain is my indebtedness to two people who kept the wheels turning, engine running and the way ahead in sight while I watched out for animals crossing.

The thing is, I don't use sat nav for I am afraid of technological innovations the way other people are wary of dirt or the sight of blood. So I relied on others more worldly than myself, namely my older brother Christopher – only fifteen months older, but far more mature throughout our very different lives, and ex-teaching colleague Helen Scally – typist extraordinaire, secretary, *manager* (would-be life coach). My boundless gratitude to both for not giving up on my waywardness.

Thanks to Nicholas Pearson at John Murray for his enthusiasm and helpful suggestions.

Appendix 1: Badger and Fox Names

	Badger names	**Fox names**
international name	Eurasian badger	red fox
Latin	*Meles meles*	*Vulpes vulpes*
trinomial (subspecies)	*M. meles meles*	*V. vulpes crucigera*
formerly	*Ursus meles, Meles taxus*	*Canis vulpes*
old names	brock, pate, grey, bawson	tod (Scot.)
other languages	blaireau (Fr.), Dachs (Ger.)	renard (Fr.), Fuchs (Ger.), zorro (Sp.), kitsune (Jap.)
derogatory	Billy	Reynard
family	boar, sow, cub	dog, vixen, cub
scientific family	Mustelidae, mustelid	Canidae, canid
adjective	meline	vulpine
collective noun	cete	skulk
den	sett	earth
dung (archaic)	werderobe	waggying

Appendix 2: Comparative Average Weights
of Adult Badger Casualties (ABCs),
Winter 2019/20 to Autumn 2022

The number in brackets after the gender is the sample size. The weights in brackets after the average weight show the range. The difference between the sexes (always in favour of the male) is shown after the curly bracket. The final figures show the weight gain or loss from one season to the next.

December 2019 – February 2020 (winter)
male (15): 11.9 kg (8.25–16.25 kg) } 1.9 kg   -0.4 kg
female (4): 10.0 kg (8.0–11.75 kg)   -1.3 kg

March–May 2020 (spring)
male (10): 10.5 kg (9.1–13.2 kg) } 0.5 kg   -1.4 kg
female (11): 10.0 kg (7.0–12.5 kg)   0

June–August 2020 (summer)
male (8): 11.0 kg (9.5–13.2 kg) } 1.3 kg   +0.5 kg
female (6): 9.7 kg (7.1–12.4 kg)   -0.3 kg

September–November 2020 (autumn)
male (8): 12.9 kg (9.6–17.0 kg) } 0.3 kg   +1.9 kg
female (9): 12.6 kg (9.0–16.5 kg)   +2.9 kg

December 2020 – February 2021 (winter)
male (7): 11.5 kg (9.5–13.5 kg)   }   −1.4 kg
female (0)   }   ——

March–May 2021 (spring)
male (14): 10.7 kg (9.1–13.75 kg) } 1.7 kg   −0.8 kg
female (9): 9.0 kg (7.0–11.6 kg) }   ——

June–August 2021 (summer)
male (19): 11.4 kg (9.0–13.0 kg) } 2.3 kg   +0.7 kg
female (7): 9.1 kg (7.2–10.0 kg) }   +0.1 kg

September–November 2021 (autumn)
male (12): 13.1 kg (8.0–16.5 kg) } 1.5 kg   +1.7 kg
female (11): 11.6 kg (7.7–14.5 kg) }   +2.5 kg

December 2021 – February 2022 (winter)
male (13): 12.4 kg (9.75–15.3 kg) } 2.1 kg   −0.7 kg
female (13): 10.3 kg (6.25–14.5 kg) }   −1.3 kg

March–May 2022 (spring)
male (23): 11.3 kg (8.2–15.0 kg) } 1.5 kg   −1.1 kg
female (26): 9.8 kg (6.8–16.0 kg) }   −0.5 kg

June–August 2022 (summer)
male (14): 11.3 kg (8.9–15.25 kg) } 1.0 kg   0
female (8): 10.3 kg (6.9–12.6 kg) }   +0.5 kg

September–November 2022 (autumn)
male (11): 12.5 kg (9.4–16.6 kg) } 1.9 kg   +1.2 kg
female (10): 10.6 kg (7.6–13.2 kg) }   +0.3 kg

Appendix 3: Reasons for Dealing with Badger Roadkill Promptly

- It might still be alive!
- It might be an illegally killed badger passed off as roadkill – only an experienced examiner can tell the difference.
- If the deceased badger proves to be a lactating female, there is always the faint chance (worth pursuing) that orphaned cubs can be rescued.
- Roadkill provides an invaluable insight into the location of badger pathways and their setts.
- It is important to remove corpses quickly so that these locations are not revealed to would-be persecutors.
- The public at large, or at least its wildlife-friendly members, are reassured when dead animals are attended to by a caring organisation; dead badgers deserve respect.
- It is imperative that corpses be returned to natural ecosystems for the feeding opportunities and nutrients they provide.
- Examining corpses is one of the most effective ways of learning about unapproachable animals at first-hand; the sooner a corpse is examined the better its condition will be.

Appendix 4: A Roe Deer Glossary

browse (v): to feed on vegetation with head up (cf. graze)
brush (n): preputial tuft of hair (buck; cf. tush)
cast (adj): of antlers shed naturally at end of breeding season
clean (adj): of new antlers when velvet shed
creep (n): worn underpass beneath fence
crotties (n): piles of droppings (cf. fewmets)
fewmets (n): individual droppings
fray (v): to strip bark from young trees with antlers
game trail (n): path though undergrowth created by the
 regular passage of wild animals including deer
gorget (n): pale throat patch in winter coat
graze (v): to feed on grass or herbs with head down (cf. browse)
kid (n): fawn
lair (n): bedding-down place
metatarsal gland (n): visible (dark) scenting area on lower
 hindleg
pearling (n): bony tubercles on antlers
pedicle (n): short bony stalk from which an antler grows
pelage (n): coat
rumen (n): first and largest of four stomach compartments
rut (n): mating season
slots (n): hoofprints
target (n): pale rump patch
tush (n): anal tuft of hair (doe; cf. brush)
velvet (n): furred skin covering developing antlers

Illustration Sources

Chapter 1 Sycamore: iStock/Hein Nouwens.
Chapter 2 Badger skull: Alamy Stock Photo/Quagga Media.
Chapter 3 Bottle: Shutterstock.com.
Chapter 4 Blowfly: Alamy Stock Photo.
Chapter 5 Fox: Alamy Stock Photo/BigJoy.
Chapter 6 Nettle: Shutterstock.com/Natalya Levish.
Chapter 7 Carrion crow: Shutterstock.com.
Chapter 8 Kestrel: Alamy Stock Photo/Walter Cicchetti.
Chapter 9 Roe deer: Shutterstock.com/Hein Nouwens.
Chapter 10 *Dead deer and Raven*, illustration from Archibald Thorburn's *British Mammals* Vol II, 1921.
Chapter 11 Ivy: Alamy Stock Photo/Quagga Media.
Chapter 12 Badger footprint: Shutterstock.com/Ariros.
Chapter 13 Otter: Alamy Stock Photo/Quagga Media.
Chapter 14 Badgers: Alamy Stock Photo/Walter Cicchetti.

Notes and Sources

The title of this book was suggested by 'Interruption to a Journey' by Norman MacCaig, about an incident involving a road-killed hare. The poem can be found on p. 8 of *Three Scottish Poets* (published by Canongate, 1992; reprinted 1998). The quote from *Lolita* by Vladimir Nabokov (first published in 1955) is taken from p. 308 of Penguin's 1959/2015 edition.

Chapter 1: The Clough

My edition of *Chambers* is the 1988 *Chambers English Dictionary* published by W. & R. Chambers Ltd and Cambridge University Press: 'a dirty, stinking fellow', p. 178.

Chapter 2: Under the Skin

The Badger: A Monograph by Alfred E. Pease (1898) was reprinted by Read Books in 2004; the gamekeeper's story is related on p. 35. 'Meles Vulgaris' by Patrick Boyle (1965) has been anthologised a number of times; see *The Penguin Book of Irish Short Stories*, edited by Benedict Kiely (1981; reprinted 2011), pp. 329–47, at pp. 342–3. Methods of cleaning skulls and bones, including maceration, are described in 'The Naturalist at Home' section of *The Amateur Naturalist* by Gerald Durrell with Lee Durrell (Dorling Kindersley, 1982), p. 273.

Chapter 3: Underground

Notes on cub development are timetabled on p. 167 of *Wild Fox* by Roger Burrows (Pan, 1968; reprinted 1973).

Chapter 4: The Compound

The Natural History of Badgers by Ernest Neal was published by Croom Helm in 1986; a graph showing male and female road casualty numbers month by month appears on p. 193. Life histories of the Calliphoridae insect family are described in *Flies of the British Isles* by Charles Colyer and Cyril Hammond (Frederick Warne & Co., 1951/1968), pp. 274–87. A photograph of *Creophilus maxillosus* appears on p. 115 of the *Collins Complete Guide to British Insects* by Michael Chinery (2005) (with accompanying text on p. 114). For the baculum bone see Neal, *Natural History*, pp. 30–1; and also pp. 19 and 62 of *Badgers* by Michael Clark (Whittet Books, 1988/1992). For hoarding see the *Oxford Dictionary of Animal Behaviour* by David McFarland (2006), pp. 96–7.

Chapter 5: Borderlands

The film *Eden Lake* (2008) was directed by James Watkins. Poets Paul Farley and Michael Symmons Roberts refer to the 'urban badlands' in their joint venture *Edgelands: Journeys into England's True Wilderness* (Vintage, 2011/12) which also put me on to the poem 'Humbles' (see chapter 10). *Running with the Fox* by David Macdonald was published by Unwin Hyman in 1987. *Our Country's Wild Animals* by H. Mortimer Batten was published by T. C. and E. C. Jack in 1930: 'delightful pets', p. 71.

Chapter 6: Grass

The concept of rewilding was first popularised by George Monbiot; see his *Feral* (Penguin, 2013/14). Monbiot also helped to popularise the idea of the lynx as a historical British native and identifies it as a specialist predator of roe deer, p. 119.

Chapter 8: Food Chains

For windfall apples see Neal, *Natural History of Badgers*, p. 135; for rabbit cohabitees see for example p. 80 of the same book. By recent estimates, the field vole population is 59,900,000; the badger population is 562,000 (The Mammal Society). H. Mortimer Batten on raisins soaked in sherry is from the book *British Wild Animals* published by Odhams Press (nd [1952]), p. 84.

Chapter 9: Dead or Alive

Konrad Lorenz's comment is from *King Solomon's Ring* (Routledge, 1952/2002), pp. 180–1. The incident involving the postman is described by Frank Holmes on p. 34 of *Following the Roe* (Bartholomew, 1974).

Chapter 10: Cheese

The illustrations in *Thorburn's Mammals* (Ebury/Michael Joseph, 1974) are on pp. 11 and 89. The standard text for tracks and signs is the *Collins Guide to Animal Tracks and Signs* by Preben Bang and Preben Dahlstrøm (various editions from

1974 onwards), a great favourite with Ray Mears. A useful glossary, 'Terms of Venery', can be found in *Field Guide to British Deer*, edited by F. J. Taylor Page (Blackwell, 3rd edition 1982), pp. 67–74. 'Humbles' by Frances Leviston is the first poem in the collection *Public Dream* (Picador, 2007).

Chapter 11: Action

Care for the Wild by W. J. Jordan and John Hughes was published by Care for the Wild in 1982 and reprinted in 1988. Helen Macdonald refers to DVCs in her essay 'Deer in the Headlights' collected in *Vesper Flights* (Vintage, 2020/21) at pp. 120–31.

Chapter 12: Winterlude

'The Tinder-Box' was a literary fairy tale written by Hans Christian Andersen in 1835.

Chapter 13: New Year

Gavin Maxwell's *Ring of Bright Water* (first published in 1960 and filmed in 1969) was followed by two sequels: *The Rocks Remain* (1963) and *Raven Seek Thy Brother* (1968). Maxwell and the otter Edal were buried at Sandaig (called Camusfeàrna in the books) on the west coast of Scotland opposite Skye, site of the author's adventures. 'The Most Incredible Thing' (1872) by Hans Christian Andersen can be found in the Penguin Classics edition of his *Fairy Tales* (2005).

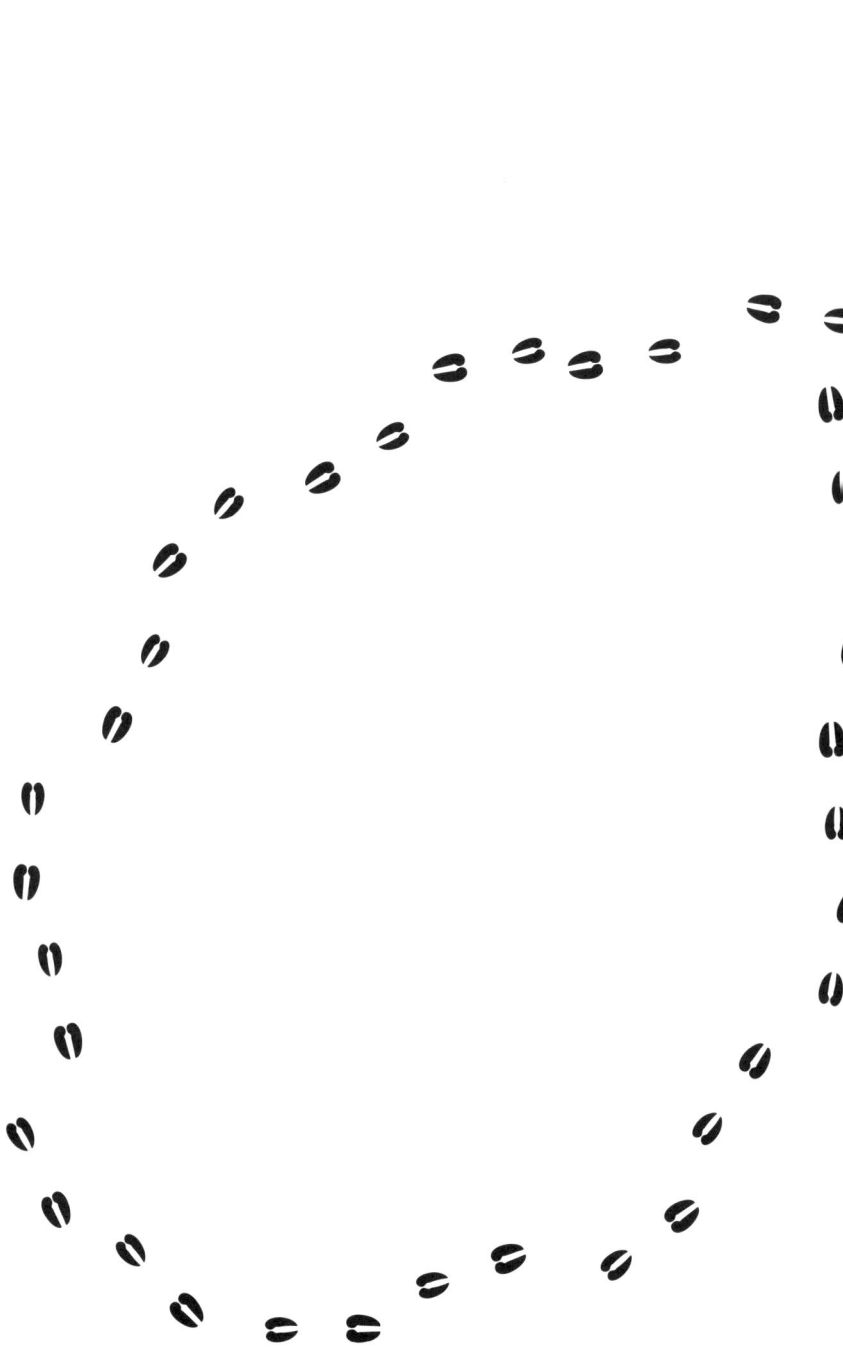